GREEN BUILDING STRATEGIES

From Plan to Profit

Jeannie Leggett Sikora

National
Association
of Home
Builders

Green Building Strategies: From Plan to Profit

BuilderBooks, a service of the National Association of Home Builders

Elizabeth M. Rich	Director, Book Publishing
Natalie C. Holmes	Book Editor
pixiedesign, llc	Cover Design
pixiedesign, llc	Composition
Sheridan Books Inc.	Printing
Gerald M. Howard	NAHB Chief Executive Officer
Mark Pursell	NAHB Senior Vice President, Expositions, Marketing and Sales
Lakisha Campbell, CAE	NAHB Vice President, Publishing and Affinity Programs

Disclaimer

This publication provides accurate information on the subject matter covered. The publisher is selling it with the understanding that the publisher is not providing legal, accounting, or other professional service. If you need legal advice or other expert assistance, obtain the services of a qualified professional experienced in the subject matter involved. Reference herein to any specific commercial products, process, or service by trade name, trademark, manufacturer, or otherwise does not necessarily constitute or imply its endorsement, recommendation, or favored status by the National Association of Home Builders. The views and opinions of the author expressed in this publication do not necessarily state or reflect those of the National Association of Home Builders, and they shall not be used to advertise or endorse a product. Although the author has made every effort to ensure that the information in this book was correct at press time, the author does not assume and hereby disclaims any liability to any reader for any loss, damage, or disruption caused by errors or omissions, whether such errors or omissions result from negligence, accident, or any other cause and is not warranting any applicable use of the material to the reader.

Printed in the United States of America

15 14 13 12   1 2 3 4 5

Library of Congress Cataloging-in-Publication Data

Sikora, Jeannie Leggett, 1969-
  Green building strategies : from plan to profit / Jeannie Leggett Sikora.
      p. cm.
  Includes bibliographical references and index.
  ISBN 978-0-86718-679-6 (alk. paper)
  1. Sustainable construction. I. Title.
  TH880.S55 2012
  690.028'6--dc23

                                        2012008481

For further information, please contact:

National Association of Home Builders
1201 15th Street, NW
Washington, DC 20005-2800
800-223-2665
Visit us online at www.BuilderBooks.com.

Cover image (right): Streetman Homes, Austin, TX

To Grandmom Vi, an admirable conservationist

CONTENTS

ABOUT THE AUTHOR

Jeannie Leggett Sikora has dedicated her career to energy efficiency. After working for several years in the agricultural sector, Ms. Sikora joined the NAHB Research Center in 1997. There, she conducted research and managed energy-related programs including the EnergyValue Housing Award and the short-lived Zero Energy Home program. She was the lead author on *Profit from Building Green*. Since 2004, Ms. Sikora has been an independent consultant, analyzing and writing about energy efficiency and renewable energy. She served as a judge for the 2012 EnergyValue Housing Awards. She holds engineering degrees from Lehigh University and Penn State. She lives in Lancaster County, Pennsylvania, with her husband and two sons.

PREFACE

When NAHB BuilderBooks asked me to write a second edition of *Profit from Building Green* (2000), I seized the opportunity to write about green building again. But I added a caveat—the book would not be a second edition. Instead, it would be an entirely new book. Although a vast field of knowledge about green building has emerged in the last decade, much of the information is scattered and obscure. I was eager to create a compendium of best practices for green building that would be of practical use to home builders. Builders have plenty to manage already—construction schedules, building inspectors, trade contractors, and home owners; I wanted to create a book that could help builders who want to "go green" do so with relative ease.

As much as possible, I have drawn from unbiased resources to get the facts and avoid marketing hype and I have combed through numerous research reports. Many of these were produced by the Building America program (http://www.buildingamerica.gov). I also relied on journal articles, manufacturers' installation instructions and technical data, and information from renowned experts in the field of *building science*. In the process of getting to the facts, I found quite a bit of misleading, incomplete, and just plain incorrect information. Therefore, I suggest that you approach information about green building cautiously. Investigate practices, companies, products, and, of course, trade contractors, before trying new methods and materials. Talk to colleagues about their experience. Enlist the help of building science consultants. Most importantly, consider how new methods and materials will affect the home's ability to be comfortable, durable, and efficient.

ACKNOWLEDGMENTS

Thank you to NAHB BuilderBooks for choosing the perfect time to ask me to write this book, to the Building America program for its long-term funding of research that has dramatically improved the body of residential building science knowledge, to organizations that excel at getting valuable building science information out to the public on the World Wide Web,[1] to the editors, and to the following individuals: Joe Wiehagen and Dave Mallay of the NAHB Research Center for sharing their knowledge; the reviewers, John Barrows, Debra Bassert, David Crump, Kevin Morrow, and Don Surrena; and Todd, Andy, and Henry for their patience, understanding, and support throughout the process. Thank you also to friends and family for your encouragement, and readers for caring about building green.

INTRODUCTION

This book was written for U.S. home builders and companies that want to improve their processes to create high-performing homes while reducing their impact on the environment. It may also be useful for home owners who want to learn about the green building process. This book addresses, for the most part, building green with conventional methods and off-the-shelf products. It is not intended for those who want details about unconventional green building methods such as straw-bale and earth-sheltered homes. It is intended to give practical information—about what green building is, how it's done, and where to get helpful tools and information to make the transition from conventional to green home building.

MANY CHOICES, NO "RIGHT ANSWER"

Which flooring is more "green," recycled carpet or bamboo? Which countertop best aligns with a project's environmental goals? Which materials should be used for the framing, roofing, windows, trim, and siding? Which lot is more suitable for a green home? In many cases, there is not one right answer. Proponents can argue the "greenness" of their preferred product or practice, but myriad considerations factor into each green building design decision.

Simply considering environmental consequences of a design or construction decision is the first step toward building green homes. Environmental consequences range from global air pollution created during a product's manufacture to local water pollution caused by erosion during the home's construction to indoor air pollutants. Each decision made during the construction process has associated environmental consequences. A green home builder weighs each of these factors during the design process.

For example, when evaluating products for a green home, you may want to consider the following questions:

- How will the product or process contribute to the home's performance?
- Are materials produced locally?
- How much energy is consumed in manufacturing the product?
- Do the raw materials contain recycled content?
- Will the material off-gas chemicals into the home after installation?
- Is the product recyclable at the end of its useful life?
- What does the product or installation process cost?

The sum of these many decisions, considered within the confines of the project's goals, budget, and builder and consumer preferences, can result in a final product that performs better than a conventionally designed and built home. Green construction considers comprehensive environmental issues such as impact on the indoor environment and emissions from the manufacture and transport of materials to a jobsite. It strives for a final product that uses fewer resources in its construction, consumes less energy for heating and cooling, and has a lower overall impact on the environment.

HOW TO USE THIS BOOK

Keeping up with the rapidly changing field of building science is daunting. The overriding goal of this book is to provide comprehensive, easy-to-read guidance to apply in your daily decision making when designing and building homes. Conferences and workshops are great, but they are time consuming and expensive. Sorting through volumes of information on the Internet is overwhelming when you have construction schedules, staff, trade contractors, and customers to manage. This book has condensed current information about green building from credible sources into a format you can easily digest.

That said, this book is not a reference work. It is not a dictionary or an encyclopedia of green building. It was designed to be read cover to cover, in the order it was written, because you must approach green building holistically and systematically. Although the book may serve as a reference, it has a beginning, a middle, and an end. It focuses on taking a home from design to finishing. Instead of implementing green home building piecemeal, wise builders understand the big picture and, thus, can implement practical approaches to building green. Although flashy finishes may make a home a showpiece, building green demands using common sense and understanding the big environmental picture. For example, saving a beautiful willow tree on a lot can add value and customer appeal that you can't buy.

As you incorporate green materials and methods, remember that construction practices change over time. What is state of the art today in building science may become archaic as new materials and methods emerge. Also, when you experiment with a new system or material, consider how it may affect other aspects of the home. For example, the movement of air, water vapor, and liquid water affect a building's durability. Carefully consider these principles when evaluating new materials and methods. Until building codes catch up with building science, there isn't a prescriptive method for building *high performance homes*. In some cases, you may wish to hire a green building consultant or an engineer who understands the implications of design decisions. These professionals can help guide you toward the best building systems for a particular climate or situation.

1

WHAT IS GREEN BUILDING?

Although green building may conjure images of solar panels, rainwater collectors, and native landscaping, a green home can take many forms and its green features may not be so obvious. In fact, a green home can look exactly like a conventionally constructed home. Fortunately, the National Green Building Standard (NGBS) certification program provides an objective American National Standards Institute (ANSI) standard for assessing and verifying various levels of green building. The standard addresses comprehensive elements of green home building from site development to home owner follow-up. Home builders who want a thorough grounding in green building principles and practices should read both the *National Green Building Standard* and the companion *National Green Building Standard Commentary* published by NAHB BuilderBooks.[2] LEED for Homes is another green certification program that comprehensively addresses the green home building process. Although LEED for Homes is not a national standard, it is a well-recognized and respected program that provides a method for defining and quantifying green homes. The bottom line is that a green building performs better than a conventional home without green features because it consumes fewer resources and lasts longer. Careful execution of building science principles helps green homes perform well.

BUILDING SCIENCE BASICS

Building science is the study of buildings and how they function. Building science principles enhance the design and operation of a home and govern materials use and installation. Applying building science correctly helps reduce a home's impact on the surrounding environment and makes the home durable, and energy and resource efficient.

Building scientists study the following areas:

- Building structural systems
- Air and moisture movement in buildings
- Energy consumption
- Building materials, their properties, and their interactions
- People and buildings (e.g., comfort, how occupancy affects energy consumption)

Scientists who study green buildings examine materials, construction methods, and systems that improve a building's environmental performance. They analyze building components and the building site, and they systematically evaluate a structure's performance in several areas. Green home construction strives for exceptional building performance while applying these guiding principles:

- Natural resource conservation
- Building durability
- Indoor air quality (IAQ)
- Resource efficiency
- Impact on the global environment

Building scientists strive to implement a systems approach to design, construction, and operation of a green home to have a low impact on the environment and still meet the needs of its occupants.

GETTING STARTED WITH GREEN

With all of the information available on green building, shifting from conventional to green home building can seem overwhelming. Fortunately, an array of resources is available to help builders. For example, the Certified Green Professional™, or CGP, (http://tinyurl.com/nahbcgp) is a professional designation builders, remodelers, and other industry professionals can earn by learning how to incorporate green building principles into homes. Class work leading to the designation provides a solid background in green building methods, as well as the tools to reach consumers. The CGP web page includes a searchable directory of builders, remodelers, and others who have earned the designation.

In addition, you should know that green building is not an all-or-nothing proposition. You can logically move step-by-step to make green improvements. An excellent starting point is to tackle energy efficiency. A large portion of the points awarded in green building certification and programs relate to energy efficiency, which has concrete benefits for home owners. Although it may not be flashy, energy efficiency is practical. Moreover, relatively small adjustments in material usage and construction techniques can reap enormous energy conservation benefits. Be cautious, however, when implementing energy-efficiency measures. For example, tightening up construction without introducing ventilation can negatively impact IAQ.

Moisture control and building durability are often linked to energy efficiency. By focusing on these areas, you can take common-sense measures to enhance a home's longevity and mitigate the troublesome and potentially litigious problems of mold and rot.

After mastering these areas, you can begin to focus on the aspects of green that may be more readily visible to home buyers, such as landscaping, finishes and materials, and then global environmental issues. Addressing interior finishes and environmental impacts beyond the walls of a home tend to be the more expensive aspects of green building, yet finishes are the most visual element of a green home and can provide a "wow" factor that requires no technical explanation to sell. And although global environmental issues may not directly impact a home owner's bottom line, environmentally minded green home customers will appreciate that you pay attention to these issues.

The United States has made strides in residential energy efficiency, according to U.S. Department of Energy (DOE) data. Average household energy consumption dropped by 31% between 1978 and 2005,[3] while home electronics usage and home sizes were growing. Despite these efficiency gains, there is plenty of room for improvement!

GREEN AND ENERGY EFFICIENT BUILDING PROGRAMS

Many national, state, and local green building programs have emerged to create a framework for evaluating and certifying green homes. In effect, these programs quantify the "greenness" of new homes—

Go to http://www.greenapprovedproducts.com for products approved to earn points to become Green Certified by the NAHB Research Center.

allowing market distinction and providing a way to independently verify whether or not a home can be considered green. Although each program is unique, all of the credible programs evaluate a green building comprehensively, considering all aspects of the home and its impact on the environment. Instead of focusing on the building itself, a comprehensive analysis examines myriad issues such as lot selection, lot development, water usage, energy and resource efficiency, IAQ, and the transfer of information to the home owner who will operate the home.

With the increasing popularity of green building rating and certification programs and the NGBS, manufacturers have made it easier to get credit for using their products. Most manufacturers' websites include a direct link to information about the sustainable aspects of their products. Many also include easily accessible information about the credits for which their products qualify under the major green building programs. Although you (or your green home verifier or consultant) must determine how various products meet green building program criteria, manufacturers are making the process easier. You may be surprised that many products and practices you currently use will qualify for points in green building rating programs—because they are produced locally, use recycled content, use fewer resources than a comparable product, are recyclable, or contain no noxious chemicals that may off-gas after installation.

Find a green building program in your area at http://www.nahbgreen.org/Resources/findlocalgreenprogram.aspx or at the U.S. Green Building Council's online *Guide to Local Green Building Programs*, http://www.greenhomeguide.com/programs.

In addition to products you may already use, many practices you employ may also earn points in the rating systems. In many cases, these best practices simply need to be added to the house plans (e.g., flashing installation details) or otherwise documented (e.g., HVAC start-up procedure) to garner credit.

A basic description of the major national green building rating programs follows.

National Green Building Standard

The NGBS, approved by ANSI, applies to single-family and multifamily new home construction, renovations, and additions (fig. 1.1).

A home can be Green Certified by the NAHB Research Center at one of four levels: Bronze, Silver, Gold, or Emerald. Many builders like the program in part because of the many options the four levels offer. Although certain core practices are mandatory for achieving certification, such as providing home owner education and including a home owner manual, the standard empowers builders to choose among many construction methods and still attain certification. Projects accrue points within seven major categories:

FIGURE 1.1 | NGBS logo

The National Green Building Standard has four levels of certification for green homes. *(NAHB Research Center)*

1. Site design and development
2. Lot design, preparation, and development
3. Resource efficiency
4. Energy efficiency
5. Water efficiency
6. Indoor environmental control
7. Operation, education, and maintenance

Under each category, the project must meet established minimums (the value depends on the desired certification level) to be certified. An independent, accredited green verifier inspects the home and submits the paperwork for certification by the NAHB Research Center. To learn more about NGBS or to locate an accredited verifier, visit http://www.nahbgreen.org.

LEED for Homes

LEED for Homes is a voluntary certification program for new single-family and multifamily green homes. Similar to the NGBS process, a third-party rater evaluates a home in eight categories of green construction: design, location, site work, water efficiency, energy efficiency, resource efficiency, indoor environment, and awareness and

In January 2010, AIA Cincinnati, a chapter of the American Institute of Architects, completed a study comparing LEED for Homes and the NGBS. This report is on the website of AIA Cincinnati (http://www.aiacincinnati.org/community/LEED_NAHB_Final.pdf). The Home Builders Association of Greater Chicago commissioned a study comparing the costs of LEED and NGBS. The results are at http://www.nahbgreen.org/Content/pdf/UrbanGreenBuildingRatingSystemsCostComparison.pdf.

education. Based on points accrued, the U.S. Green Building Council certifies a home at one of four levels: Certified, Silver, Gold, or Platinum. All projects must include 18 practices which are assigned no point value. Points are awarded for additional practices selected by the design team.

Environments for Living Certified Green Homes

The Environments for Living (EFL) Certified Green Home program is designed to help builders construct energy-efficient homes with an added green element. It is an outgrowth of the longstanding EFL program focusing on energy efficiency. The program addresses only the home (not site design or development). Masco Home Services, which installs many building-science-related products, operates the program. EFL's certification program relies on most of the same building science principles as the other green building certification programs and also offers a heating and cooling cost guarantee on certified homes. This guarantee can be an effective marketing asset for builders. It is one of many marketing tools offered to companies that have homes certified through the program.

ENERGY STAR

This commonly recognized label can be applied to homes as well as to the appliances and electronics that often carry it. Requirements for earning the U.S. Environmental Protection Agency's (EPA) ENERGY STAR label have increased since the program began in 1995. Version 3 ENERGY STAR program guidelines that were being phased in as this book was written include substantial changes to reflect building code changes and building science research. Most notably, these guidelines augment existing inspections with additional inspection points. They also place limits on duct leakage.

DOE Challenge Home

Participants in this voluntary U.S. DOE program (http://tinyurl.com/BuildersChallenge), which was being updated to Version 2 at press time, are challenged to go beyond ENERGY STAR requirements by incorporating mandatory practices to enhance durability, use water efficiently, reduce the source of indoor pollutants, and provide rough-in mechanical systems for future on-site renewable energy production.

Companies that participate in the DOE Challenge Home can leverage their marketing dollars with turnkey marketing materials including bright yard signs, colorful brochures, and rating stickers to mount on a home's electrical panel.

Energy Cost Guarantees

Energy cost guarantee programs (or energy use guarantees) can help you market your homes. With these programs, builders (or a third party such as a utility) warrant that heating and cooling costs will not exceed a specified amount. In a typical guarantee, builders pay a fee for the service, which includes house plan review, on-site inspections, home performance testing, and computer model generation to predict the home's energy use. With current utility cost data and other parameters (e.g., thermostat settings), the software calculates how much the home will cost to heat and cool. If bills exceed this guaranteed cost and home owners have operated the home within the constraints of the guarantee, the program (or the builder) reimburses the home owner for the excess. Current energy cost guarantee programs include Environments for Living, ComfortHome, SystemVision Home Program Comfort and Energy Use Guarantee, and Tucson Electric Power's Guarantee Home program.

ENERGY STAR 2.5 AND 3 GUIDEBOOKS

Go online to http://tinyurl.com/ENERGYSTARCHECKLISTS and scroll down to the Inspection Checklist Guidebooks section to download the following illustrated guides that explain best practices for energy efficiency:

- *Building Science Introduction*
- *Thermal Enclosure System Rater Checklist Guidebook*
- *HVAC System Quality Installation Rater Checklist Guidebook*
- *Water Management System Builder Checklist Guidebook*

Regional Programs

In addition to the major national green building programs, there are several well-established regional programs. Most are offered through a utility, home builders association (HBA), government agency, or nonprofit organization. Examples include EarthCraft in the Southeast, Green Built Home (Wisconsin), Green Built Michigan, Green Built Texas, Earth Advantage in the Pacific Northwest, and Built Green Colorado. Popular regional programs can add value to your green marketing because consumers in the local market recognize them. In addition, many programs offer checklists, participant directories, and other resources for participating builders.

THE PRICE OF GOING GREEN

Many green products and practices cost roughly as much as conventional products and methods; others cost more and some can cost less. For example, orienting a house for solar or installing windows using best practices does not cost more but careful air sealing probably will incur a few hundred extra dollars in labor and materials. Likewise, recycled-content countertops will cost more than laminates but the cost may compare favorably with granite. Tight construction reduces loads so a home can be heated and cooled efficiently with smaller heating and cooling equipment, which saves money. By simply adjusting techniques to adhere to best practices or by getting information about a material's green properties from suppliers and manufacturers, you are going a long way toward green building certification without adding to construction cost.

The incremental cost of building green homes is difficult to pin down. Based on 16 case studies, the 2005 report *Costs and Benefits of Green Affordable Housing* found the average total cost premium to go green with affordable housing was 2.42%.[4] Another 2008 study compared a code-minimum home in two locations to similar NGBS and LEED for Homes certified homes.[5] The study found the cost of compliance (excluding certification fees) for NGBS ranged from 1.1% to 16.9% and LEED for Homes, 3.6%–22.9%, of the code-compliant home's cost. Although these results certainly are not definitive and will change as certification requirements and codes change, the examples demonstrate that green does not necessarily equal expensive. On the other hand, luxury homes built with more elastic budgets may incur much higher premiums, if environmentally motivated customers weigh ecology—and the "wow" factor of expensive finishes—more heavily than economics.

To gain recognition for your green building efforts and to market your homes as certified green, you will need to pay certification and verification fees. The NGBS-certifying agency, the NAHB Research Center, charges $200–$500 to certify a single-family home. This fee does not include the cost of verification, which is independently negotiated with an NAHB Research Center-accredited verifier. EFL charges about $600 for program fees including the cost of providing an energy cost guarantee on the home. LEED for Homes program fees are $375–$525 for single-family homes. DOE Challenge Home and ENERGY STAR don't charge program fees but you will incur costs for Home Energy Rating and other documentation. DOE Challenge Home and ENERGY STAR are aligned so you can minimize or eliminate redundant costs.

ESTIMATING COSTS

To estimate the cost of various energy-saving technologies, such as *tankless water heaters* and high-efficiency furnaces, you can use the Residential Energy Efficiency Measures Database (http://www.nrel.gov/ap/retrofits/index.cfm) created by the National Renewable Energy Lab. The lab created the database for remodelers but much of the information it contains applies to new homes too.

GOING GREEN WITHOUT APPREHENSION

You can improve your homes to be more durable, to use less water and energy, and consume fewer materials. If you are already using current best practices in residential construction, you probably won't have to dramatically change what you are doing to become a green builder: you will already be accustomed to using many of the materials and methods described in this book. Many small steps can add up to dramatically reduce a home's environmental footprint. You do not need to be a pioneer; other builders have done it. Experts in energy and resource conservation and reuse and product manufacturers have developed materials, resources, and best practices to follow.

In general, however, going green requires a bit more advanced planning and careful decision making. Fortunately, building scientists, home builders, and manufacturers are continuously developing better ways to make building, and buildings, more environmentally friendly. Moreover, national and regional green building programs provide information to help you make the best decisions possible for your green home construction projects. Green building offers many challenges and rewards. The biggest challenge may be deciding to make the leap from traditional construction to green building.

2

GREEN HOME DESIGN

Successful green home construction demands design and careful planning. Begin by establishing a project team with the power to set goals, create a budget, and oversee the entire project. This team may include the general contractor, trade contractors, design professionals, sales and marketing personnel, suppliers, third-party raters, green home consultants, and others. Green home performance depends upon various and distinct components of a home working harmoniously. Therefore, incorporate the entire design and construction team, including trades and vendors, to develop practical and effective solutions.

LOCATION AND LAND DEVELOPMENT

The green home design process begins well before a house plan is drawn. One of the first green design considerations is a home's location, which can greatly influence the project's environmental impact. The following are some considerations:

- Is the lot environmentally sensitive (e.g., does it contain *steep slopes* that would be disturbed by development, or is it prime agricultural land, wetlands, or wildlife habitat)?

- Is the lot in an already developed area that is served by public water, sewer, and roads?

FINDING A GREEN BUILDING CONSULTANT

Certification programs can help builders locate professionals who have received training and demonstrated knowledge of green building programs and practices. NGBS Accredited Verifiers and LEED AP Homes professionals demonstrate proficiency within their respective certification programs. Building Performance Institute (www.bpi.org) offers several certifications for individuals with demonstrated expertise about specific home performance topics.

- Is the lot in close proximity to public transportation and services?

Answering these questions and others when selecting a homesite can help you and the home owner choose a lot that will support the goals of a green building project.

To improve its environmental performance, a green home may be located on

- a previously developed site (e.g., a lot with an abandoned structure);
- an EPA-recognized brownfield;
- an infill lot within an existing development that is served by roads and utilities;
- a lot near public transportation; or
- a lot within walking distance to community services.

PRIME FARMLAND

Green building certification programs such as NGBS award points for avoiding development on areas classified as Prime Farmland according to the NRCS Web Soil Survey (http://www.websoilsurvey.nrcs.usda.gov). To find this classification, you can use the online Web Soil Survey. After defining your Area of Interest, click the Soil Data Explorer tab and then Suitability and Limitations for Use. Choose Land Classifications and then Farmland Classifications to determine whether a parcel is Prime Farmland. The Web Soil Survey also includes other valuable information about a soil's suitability for residential development.

To minimize impact, development should avoid disturbing wetlands, steep slopes (generally defined as average slope greater than 15% or 20%,[6] prime agricultural land (as defined by the United States Department of Agriculture's (USDA) Natural Resources Conservation Service (NRCS), or areas of critical habitat for endangered species such as those identified by the U.S. Fish and Wildlife Service.

The U.S. Fish and Wildlife Services online mapping tool (http://www.crithab.fws.gov) and your state's department of natural resources have information about critical habitat. States often have tougher restrictions than the federal regulations.

The key to green development is not only to avoid environmentally sensitive areas, but also to carefully design a site to minimize disruption of sensitive natural resources. For example, wetlands can be restored or open space preserved as a wildlife corridor. As the supply of land for development dwindles, it becomes increasingly important for builders and developers to consider how construction will impact natural resources and how to minimize that impact.

TAKING STOCK OF NATURAL RESOURCES

Once you identify a possible location for a green home, the next step is to identify existing natural resources, including water, vegetation, and soil, and develop plans to preserve each one during construction:

Water. Land development that considers the natural flow of water will improve lot drainage and can protect seasonal ponds or other features that provide habitat for fish, birds, and other wildlife.

Vegetation. Too often builders and developers do not consider existing trees when they position a home on a lot or commence construction, even though these trees provide natural beauty and shade.

Soil. Your site planner must consider the specific characteristics of soil types to plan drainage strategies and determine where to place the home on the lot.

Topography. Work with the topography rather than against it. You will reduce excavation and fill costs and minimize the loss of valuable soil.

Sun. The most energy-efficient homes use the sun's heat and light as "free" energy sources. It is wise to plan ahead for how a home will access the sun even after trees have matured. The sun is a valuable resource for naturally heating and lighting a home. It can be used to supplement mechanical heating in a *sun-tempered design* or it can provide most or all of the home's warmth in a *passive solar* design.

Taking stock of the available natural resources and creating a plan for conserving these resources throughout construction can prevent the destruction of irreplaceable natural resources and add value and beauty to the home.

Water

In designing a green homesite, keep the flow of water as natural as possible. This protects the natural environment by allowing rainwater to percolate back into the ground, rather than being swept away in storm drains. When the home is also sited correctly, water will be diverted away from the structure to protect its foundation.

To maintain the natural flow of water on a lot

- avoid altering the natural topography and water channels;

- limit impervious surface area created by the house, driveway, patios, and sidewalks (by making the home smaller, building two stories instead of one, using permeable paving surfaces for hardscapes, and creating driveway strips instead of fully paving a driveway, for example); and

- create areas to capture rainwater, such as *rain gardens, vegetated swales*, and *rainwater catchment systems*. These systems reduce runoff by retaining storm water until it percolates back into the ground. Preventing runoff reduces erosion, water pollution, and flooding, and recharges groundwater.

The Practice of Low-Impact Development (search for the title at http://www.huduser.org) is a free comprehensive resource for planning and designing a green home or community. In addition, the website of the Low Impact Development Center (http://www.lowimpactdevelopment.org) offers sample designs for rain gardens and other resources.

Vegetation

Trees, forested areas, and vegetation around streams are attractive to home buyers. They also protect soil from erosion, filter storm water runoff, and help moderate temperatures in and around a home. For example, trees on the western side of a home shade it from the afternoon sun.

In the early planning stages for a green home, a builder or other knowledgeable professional should inventory and develop a plan for saving valuable vegetation. A certified arborist can advise you about tree care during construction, including pruning and feeding strategies to keep existing trees healthy. Builders and developers also incorporate easily recognizable "tree-save" areas that protect trees from potential damage during site development and construction. These organizations can help you find qualified tree-care professionals:

- The International Society of Arborists (http://www.isa-arbor. com/faca/findArborist.aspx) maintains an online database of its certified arborists.

- The Tree Care Industry Association (http://www.tcia.org) includes a searchable database of accredited member businesses.

- The American Consulting Arborist Association (http://www. asca-consultants.org/find/index.cfm) lists Registered Consulting Arborists.

Soil

General soil surveys, such as those from the NRCS, help you determine whether a site is suitable for development. For example, these surveys determine if *hydric soils* exist. Developers should avoid sites with these moisture-laden soils because they indicate the area is a permanent wetland, a location that periodically floods, or a place where moisture ponds. If the soil is deemed suitable for development, a soils test (conducted by the local Cooperative Extension Office) can determine the soil's needs for reestablishing vegetation. This inexpensive test can determine what is needed (e.g., nitrogen, organic matter, increased permeability, pH balancing) for vegetative growth.

Once you settle on a lot, the soil investigation is not finished. It is necessary to understand each particular site's soil characteristics to develop the site properly for a green home, including planning for drainage. A thorough investigation of the soil will help you or the site designer determine where to place buildings, driveways, drainage areas, and other features.

Topography

When you capitalize on a site's natural topography, rather than trying to fight it, you can save money because you will reduce cutting and filling. For example, when a lot or parcel includes steep slopes, you should avoid disturbing them. If you can't avoid developing steep slopes, hire a soil mechanics engineer to conduct a *soil stability study* to help inform site design. Although the first priority of a soil stability survey is to prevent a landslide and to protect the building and its occupants, it can also help you obtain a design that prevents erosion and keeps subsurface and surface water away from the building. Engineers can also evaluate soil to glean information about the water table.

Although building on steep slopes is not ideal, with a properly engineered design, developers can minimize the risk of structural damage and erosion by using retaining walls, terraces, vegetation, or other soil stabilization techniques.

Solar Orientation

Any home in any climate can benefit from the sun's warming effects and its ever-present natural light. Instead of routinely orienting a home according to a street layout, place it so the design can maximize the use of solar energy. Although not all home designs can be passive solar, you can tweak most of them to be sun-tempered, or partially heated by the sun. All homes can be designed to use the sun's natural light.

Passive Solar Design

Passive solar design incorporates architectural features, materials, and a means to circulate the sun's warmth to allow a home to rely primarily on the sun for heating. For example, a passive solar design might include south-facing window area with overhangs or more sophisticated controls to shade the windows from excessive sun, dark floor tiles to absorb and store solar heat, and a system for distributing this stored energy to other areas of the home.

Passive solar design is not simply placing a wall of windows on the south side of the house. In fact, this misconception often leads to overheating a space or, conversely, placing excessive demand on the home heating system because of heat loss through the windows.

Unlike active solar energy systems such as solar water heaters, passive solar design typically includes few moving parts. However, a professional experienced in solar design must perform complex, location-specific calculations of window area, overhang width, and degree of thermal storage. Passive solar design requires careful consideration of the local climate and solar energy resource, building orientation, and landscape features, to name a few.

The principal elements of passive solar design (fig. 2.1) include proper building orientation, proper window sizing and placement, design of window overhangs to reduce summer heat gain and ensure winter heat gain, and proper sizing of thermal energy storage mass (for example a brick wall or masonry tiles). In a passive solar home, the heat is distributed primarily by natural convection and radiation, although fans can also be used to circulate room air and ensure proper ventilation.

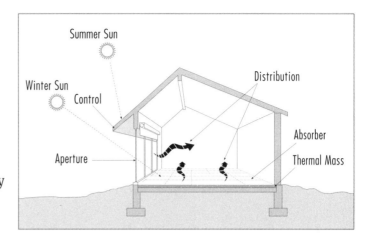

FIGURE 2.1 | **Passive solar design elements**

Elements of a passive solar design include windows located for solar heat gain, shading to prevent unwanted heat gain, solar-energy-absorbing surfaces, a thermal storage device (like concrete, brick, or water), and a method for distributing stored heat. *(U.S. DOE)*

Sun-Tempered Design

In a sun-tempered design, the builder considers the climate and the sun in designing the home but does not attempt to heat the home primarily with the sun. The following strategies help moderate heating (and cooling) loads in a sun-tempered design:

- Limiting windows on west-, east-, and north-facing facades
- Orienting a home so its long axis sits within 20 degrees of east-west
- Ensuring that south-facing windows have sufficient overhangs to prevent overheating in the summer
- Limiting window area on the south face and using windows with a *solar heat gain coefficient (SHGC)* greater than 0.40
- Using tubular skylights or installing shades on traditional skylights
- Placing the house in an area that is naturally protected from winter winds
- Adding features like operable skylights to promote natural breezes throughout the home

- Shading east- and west-facing windows with covered porches, vegetative fences, awnings, or architectural features such as a detached garage, that fully shade these areas

- Using cool roofing materials or attic *radiant barriers* in warmer climates to reduce attic temperatures

For many of today's building lots and construction techniques, a sun-tempered design is more practical than passive solar homes, especially for production homes where orientation is usually constrained.

DESIGN SOFTWARE

Many software programs are available for designing green homes. These include *life cycle assessment (LCA)* and energy simulation programs.

Building for Environmental and Economic Sustainability (BEES) software, an LCA program, is free. All LCA programs evaluate the environmental performance of user-specified building materials for their impact on IAQ, their *embodied energy*, and their human toxicity, among other factors. You can use LCA to compare the environmental impact of materials, such as different flooring options. Using LCA programs usually garners credit from green building rating programs.

Energy simulation programs such as Energy-10, eQuest, EnergyGauge, and REM/Design, are software tools for designing energy-efficient homes. These programs inform decisions about insulation, window selection and sizing, air sealing measures, lighting, and appliances. Before finalizing home plans, the user enters most of the information about a building, such as window location and U-value, *R-value* of the wall system, and efficiency ratings of mechanical equipment, into the software program. These programs conduct hourly simulations of the home, using the construction details and climate specified to predict annual energy consumption. You can use them to evaluate the impact of design decisions on predicted energy usage.

The U.S. DOE maintains a Building Energy Software Tools Directory (http://apps1.eere.energy.gov/buildings/tools_directory). Many of the tools are geared for commercial buildings, but you can narrow the list of options by searching for "residential" in the program descriptions. Like the BEES LCA software, some of the programs are free.

HIGH PERFORMANCE HOME PLANS

Construction crews depend on home plans to execute a home design as intended. When you combine plans with detailed scopes of work (see Chapter 11: Quality Control and Construction Management), you will be able to track construction details that are essential to the final product's performance. Ideally, building plans for a green home include details about durability, structural design, and the integration of mechanical systems into the conditioned space of the home. For example, if plans call for a drop ceiling in a hallway to accommodate ducts in conditioned space, the plans should clearly illustrate how the chase should be constructed, including where to use sealants to prevent air infiltration from the chase into the space above. Include green design elements on the blueprints to eliminate guesswork in the field that can prompt improper installation and compromised durability.

A complete set of plans for a green home includes the following details:

Site

- A site plan that identifies the limits of clearing, grading, and excavation, and the vegetated areas that will remain undisturbed
- Specifications for driveways, patios, walkways and other hardscaping
- Soils information, including specifications for structural soils and topsoil and, where problematic soil conditions exist, a detailed soils analysis
- Specification for backfill
- Benchmark, elevations, and contours
- Storm water management plan that preserves natural drainage and mimics the natural flow of water on the lot
- Landscaping plan that preserves natural resources

Building

- Detailed framing plan to limit unnecessary lumber usage and facilitate having mechanical systems in conditioned space
- Moisture management strategy details including

- site and foundation drainage system details with specifications for the drain tile;
- specifications for foundation waterproofing, including foundation wall coating material;
- capillary breaks (such as heavy-duty *polyethylene*, dampproofing sealers, or proprietary membranes) to prevent water absorption from below the slab, between the footer and the foundation, and between the foundation wall and sill plate;
- house wrap attachment details including overlapping, taping of seams, and window and door opening installation details;
- vapor retarder specifications;
- finished grade requirements;
- proper drainage details for absorptive claddings like stucco, stone, and brick; and
- flashing details for
  - windows (fig. 2.2), skylights, and doors;
  - roof valleys;
  - deck-to-building intersections;
  - roof-to-wall or roof-to-chimney intersections;
  - sidewall penetrations (e.g., vent pipes) that depict integration with the house wrap; and
  - kickout flashing.

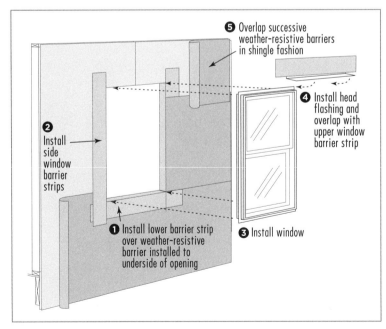

❺ Overlap successive weather-resistive barriers in shingle fashion

❷ Install side window barrier strips

❹ Install head flashing and overlap with upper window barrier strip

❶ Install lower barrier strip over weather-resistive barrier installed to underside of opening

❸ Install window

FIGURE 2.2 | Window flashing details

An example detail to be included in a set of high performance home plans. *(U.S. DOE)*

- ductwork layout performed in accordance with industry standards (ACCA Manuals J, S, and D).

- Details of the ventilation system

Building America offers Building Science Information Sheets with construction details for high performance homes (http://tinyurl.com/HPhomedetails). These details address moisture management, indoor air pollutants, and energy usage. Many details can be adapted or customized for high performance construction drawings.

TAKE YOUR TIME WITH GREEN HOME DESIGN

Handling the design phase well will help ensure that a green home meets the project's goals. Take time to hammer out critical details before you break ground. Advance planning will help you avoid roadblocks in the middle of the project or later. There are many resources available to help you become a pro at managing the design process for a high performance home.

The Energy and Environmental Building Association (http://www.eeba.org) publishes a series of climate-specific *Builder's Guides*. These are valuable tools for builders who want to revise building methods for energy efficiency and moisture control. The organization also offers Houses that Work one-day training seminars focused on building high performance homes.

3

THE GREEN HOME CONSTRUCTION PROCESS

Any builder can improve each stage of building, from planning and design to construction, to make green improvements. Fortunately, you have the freedom and creativity to prioritize what to include in building the best green home for a particular buyer. For example, improving IAQ for a chemically sensitive customer may be the overarching concern for one project. Energy efficiency and budget may be the primary concerns for another. You can balance sound building science with consumer preference and budget in constructing green homes, although environmental issues influence each stage of construction, from foundation to roofs. This chapter discusses how to improve environmental performance at each stage.

SAVING TOPSOIL

Topsoil is a valuable resource. It contains organic matter and nutrients essential for reestablishing vegetation on the site. When soil erodes, it pollutes local waterways. Therefore, treat the soil with care and try to preserve as much topsoil as possible while constructing a green home.

One way to minimize soil erosion is to keep existing vegetation intact. Although you may not be able to preserve all of it, you probably don't need to grade the entire lot. Also, you can grade in stages to minimize soil exposure at any given point. At a minimum, stake limits for clearing and grading before construction starts. Beyond that, you can incorporate vegetated buffer strips at the borders of a lot to help keep soil on the site. You can also establish temporary vegetative cover using fast-growing grasses like rye. Use a layer of compost or mulch or a commercially available erosion control blanket to reduce runoff after seeding and before the temporary vegetation is established.

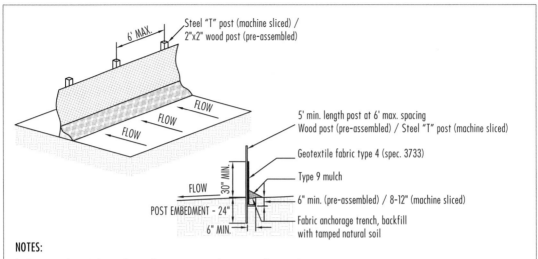

Steel "T" post (machine sliced) /
2"x2" wood post (pre-assembled)

6' MAX.

FLOW
FLOW
FLOW

FLOW

30" MIN.

POST EMBEDMENT - 24"

6" MIN.

5' min. length post at 6' max. spacing
Wood post (pre-assembled) / Steel "T" post (machine sliced)

Geotextile fabric type 4 (spec. 3733)

Type 9 mulch

6" min. (pre-assembled) / 8-12" (machine sliced)

Fabric anchorage trench, backfill
with tamped natural soil

NOTES:

1. Inspect and repair fence after each storm event and remove sediment when necessary

2. Removed sediment shall be deposited to an area that will not contribute sediment off-site and can be permanently stabilized

3. Silt fence shall be placed on slope contours to maximize ponding efficiency

4. Installation and material requirements shall be in accordance with mn/dot 3886 for the type of silt fence installed
   (pre-assembled or machine sliced)

NOT TO SCALE

FIGURE 3.1 | Diagram of a silt fence

Properly installed and maintained, a silt fence is part of a solution that prevents runoff.
(City of Moorhead, Minnesota)

Other options for controlling erosion and runoff include *silt fences* (fig. 3.1) and *compost socks* designed to trap silt. Despite their widespread use, these fences often are not installed properly or they are compromised during construction so they do not adequately contain sediment. In a proper installation, the ends of silt fences extend a foot or so uphill from the base of the fence. The bottom of a silt fence should be installed in a 6"-deep trench. Anchor the fence with soil and support it at least every 6' with wooden stakes. Inspect and clean silt fences after rainfalls of ½" or more; a clogged fence will not function properly.

Stockpile the soil scraped off during clearing and grading in a designated area and minimize the time it is exposed to the elements. You can use silt fences to contain the stockpile and establish temporary vegetation on it to minimize erosion and runoff. In addition, protect storm water inlets with sediment controls and inspect these controls after a rainfall.

PRESERVING PLANTS

Heavy equipment can damage or kill vegetation, especially trees, by compacting the soil around the root zone. Compaction purges air spaces, stunts root growth, and compromises a plant's drought resistance. A certified arborist conducting a natural resources inventory on the site can determine which trees are valuable and likely to survive the construction process. The arborist can develop a plan for preserving each tree to be saved.

Create vegetative protection areas that encompass the plants' critical root zones by erecting protective fencing or caution tape around the critical root zone (fig. 3.2). The size of the critical root zone is a subject of scientific debate. Generally, you should try to protect at least the area that falls under the tree's canopy or up to 1 ½ times the tree's height. Regardless, many contractors visiting a jobsite won't distinguish a noxious invasive weed from a treasured native plant so it is your responsibility to protect the vegetation.

Beyond avoiding their root zones, existing trees may require other special care to survive the stress of construction. The arborist's plan should include provisions for pruning, fertilizing, and watering during construction. For more information, see The North Carolina Cooperative Extension Service's (http://www.ces.ncsu.edu) fact sheet, *Construction and Tree Protection* (fig. 3.3). When hiring tree-care professionals, ensure they use the industry standard practices outlined in *Standards for Tree Care Operations* (ANSI Standard A300, http://www.tcia.org/standards/A300.htm) when pruning, fertilizing, and transplanting.

5 STEPS TO MINIMIZE EROSION

1. Stake out the limits of clearing and grading before construction.

2. Stage grading activities to reduce the area of exposed soil at any one time.

3. Maintain vegetated buffer strips at the lot's borders.

4. Establish temporary vegetation or other erosion-control measures over exposed soil, including stockpiled topsoil.

5. Properly install and maintain silt fences and sediment filtration devices.

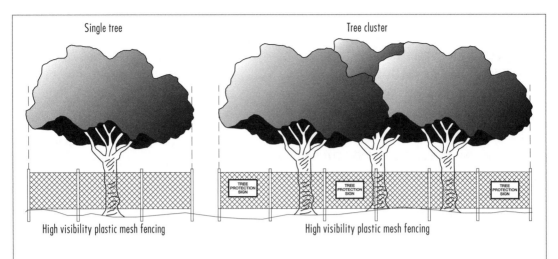

Single tree | Tree cluster

High visibility plastic mesh fencing | High visibility plastic mesh fencing

PROTECTIVE FENCING: Orange vinyl construction fencing, chain link fencing, snow fencing or other similar fencing at least four feet (4') high and supported at a maximum of ten-foot (10') intervals by approved methods sufficient enough to keep the fence upright and in place. The fencing shall be of a highly visible material, and shall have a tree protection sign affixed to the fence every twenty (20') feet in such a manner to be clearly visible to the workers on-site.

PRIOR TO CONSTRUCTION: The contractor or subcontractor shall construct and maintain, for each protected tree or group of trees on a construction site, a protective fencing which encircles the outer limits of the critical root zone of the trees to protect them from construction activity. All protective fencing shall be in place prior to commencement of any site work and remain in place until all exterior work has been completed.

NOT TO SCALE

FIGURE 3.2 | Typical tree protection fencing

Fencing is part of a strategy to help valuable trees survive construction.
(City of Southlake, Texas)

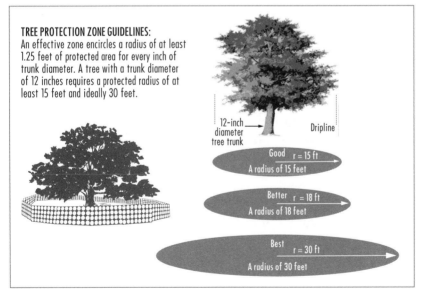

TREE PROTECTION ZONE GUIDELINES:
An effective zone encircles a radius of at least 1.25 feet of protected area for every inch of trunk diameter. A tree with a trunk diameter of 12 inches requires a protected radius of at least 15 feet and ideally 30 feet.

12-inch diameter tree trunk | Dripline

Good r = 15 ft
A radius of 15 feet

Better r = 18 ft
A radius of 18 feet

Best r = 30 ft
A radius of 30 feet

FIGURE 3.3 | Tree protection zone guidelines

Avoiding root zones and other measures can help trees and other vegetation survive construction. *(Construction and Tree Protection, North Carolina Cooperative Extension Service)*

MATERIAL STORAGE

To optimize their moisture performance and durability, keep building materials dry before installation. On-site building material storage should protect sheet goods, lumber, trusses, siding, and other moisture-sensitive products from the elements. Manufacturers often recommend storage and handling techniques. In general, store materials flat, off the ground, and covered with waterproof material such as plastic or house wrap. Sunlight can cause UV degradation of materials, so store them under a roof if possible.

REDUCING CONSTRUCTION WASTE

You can cut construction waste by 50% or more by establishing and implementing a rigorous waste management plan:

- Identify local markets for construction waste. Your local solid waste authority and the online directory at http://www.earth911.org are good resources.

- Set goals, create a plan for meeting them, and post a construction waste management plan on the jobsite.

- Create clearly labeled, conveniently located bins for recyclables.

- Provide incentives for cooperative trade contractors and disincentives for those who do not cooperate.

- Follow proper handling procedures for materials (e.g., keep drywall scraps dry).

- Provide a place for day-to-day recyclables like bottles and cans.

Jobsite materials that usually are recyclable include asphalt, brick, concrete, drywall, carpet, ceiling tiles, cardboard, paper, plastic film, polystyrene, scrap metal, wood, and yard waste. Habitat for Humanity ReStores collect undamaged construction materials for reuse.

DESIGNING AND CONSTRUCTING FOUNDATIONS

Foundation design and construction are critical in green home design not simply for structural integrity but also because the foundation forms the home's connection with the ground: it must be able to manage often-relentless subgrade moisture, prevent radon intrusion, and moderate energy usage.

The NAHB Research Center offers a series of online videos
about moisture management in residential construction
(http://www.toolbase.org. Click Best Practices/Construction Videos).
The techniques described address mixed-humid climates but many
of the strategies also apply to other climate zones.

Excavation, drainage, waterproofing, backfilling, and insulation play crucial roles in the long-term performance of a foundation. Many of the design elements presented here come from the National Green Building Standard, LEED for Homes, and Revisions to *Quality Management Products: Four Scopes of Work for High Performance Homes,*[7] which includes a sample high performance scope of work for excavation and backfill for basement foundations for the mixed-humid climate.

Excavation

When excavating for a green home, moving and removing the smallest amount of soil is likely the best way to protect the site's natural resources. Beyond that, excavators can reduce impact by driving equipment only on designated areas (or by using efficient equipment). Before excavation, the lot should be accessible for heavy equipment with a temporary driveway. Stake the house footprint clearly according to the site and architectural plans. Clearly state where the excavator should stockpile topsoil and how to protect it from the weather. Make the excavator aware of erosion control measures and his responsibility for maintaining them.

Drainage

Although waterproofing materials can help keep foundations dry, most materials cannot resist the *hydrostatic forces* of groundwater. Hydrostatic force is exerted from standing water on one side of a wall or floor. Because of the hydrostatic force, the water moves through the assembly toward the other, lower pressure side. The foundation drainage and rainwater management systems must be designed to reduce hydrostatic force by keeping water away from the foundation.

Surface runoff management is crucial to a foundation drainage system. A roof-and-gutter system that directs water away from the foundation can control surface runoff. Further, a storm water plan should capitalize on the site's topography and natural drainage features to divert rainwater from the foundation.

Interior and exterior perimeter drains that are sloped to discharge to daylight or a sump pit are the most effective systems for promoting foundation drainage. Interior perimeter drains also can be integral to a radon-reduction system. Drain tile must be code-approved material (e.g., drainable concrete forms like those shown in figure 3.4 or schedule 30 perforated PVC). For best performance, follow these steps:

- At the base of the trench, lay geotextile filter fabric around the footing perimeter.

- Lay 4" of washed stone or #1 and #2 gravel atop fabric, extending about 12" from the foundation.

- Install 4" diameter drain tile, to daylight or sealed sump pit, with the holes oriented downward and level or at a slope of ½" in 4'. To be effective, the holes must be below the top of the footing. Alternatively, you can install a drainable concrete form that stays in place after pouring the footers.

- Include a cleanout for future flushing of the system.

- Install 10" of washed gravel or #1 and #2 stone on top of the drain tile.

- Wrap geotextile fabric around all sides of the drain tile pipe and stone bed.

FIGURE 3.4 | Drainable concrete form

Workers begin assembling a drainable concrete form that remains in place after the pour to serve as part of the home's moisture management and radon control systems. *(Jeannie Leggett Sikora)*

Diverting Rainfall

Even with an excellent drainage system, a house still needs a functioning gutter and downspout system to divert water from its foundation (or a system for harvesting rainwater for reuse in the house

Innovative measures such as vegetated roofs, rainwater harvesting systems, and permeable pavement can lessen the amount of rainwater to be managed on a lot.

or for watering). To effectively manage storm water, locate the outlet of conventional gutter and downspout systems at least 5' from the foundation and slope the finished grade away from the house at least 5% (equivalent to 6" in 10').

Foundation Wall Waterproofing

Waterproofing is the next line of defense after an effective drainage and conveyance system is in place. There are many options today, unlike in the past when only tar-based dampproofing was available. Consider your own experience, anticipated performance, environmental toxicity, and local availability when choosing from among the array of available materials, products, and techniques for a waterproofing system.

Preventing Capillary Action

Just as a tree draws water upward through its roots, water can also be drawn into the foundation from below through tiny pores in the slab or foundation wall. To prevent this *capillary suction*, you must install a barrier (a capillary break) between the footer and the foundation

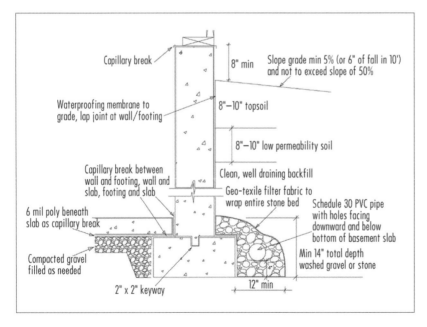

Capillary break

8" min — Slope grade min 5% (or 6" of fall in 10') and not to exceed slope of 50%

Waterproofing membrane to grade, lap joint at wall/footing

8"–10" topsoil

8"–10" low permeability soil

Capillary break between wall and footing, wall and slab, footing and slab

Clean, well draining backfill

Geo-texile filter fabric to wrap entire stone bed

Schedule 30 PVC pipe with holes facing downward and below bottom of basement slab

6 mil poly beneath slab as capillary break

Compacted gravel filled as needed

Min 14" total depth washed gravel or stone

2" x 2" keyway

12" min

FIGURE 3.5 | Moisture control system for poured concrete foundation wall

A capillary break between the footer and the foundation wall prevents the upward movement of water from the ground into the foundation wall. *(NAHB Research Center)*

wall, between the foundation wall and the sill plate, and beneath the concrete slab (fig. 3.5). Many builders use closed-cell polyethylene foam *sill sealer.* For slab foundations, extending polyethylene around the stem wall can prevent capillary suction through the wall—but you also need a sill sealer to serve as a secondary capillary break between the foundation and wall system.

You can prevent ground moisture and water vapor from seeping into the slab by placing 6-mil (minimum thickness) plastic sheeting, with seams overlapped at least 6" and taped, between the slab and the ground. The sealed plastic sheeting can perform double duty as a component of a radon mitigation system.

CAPILLARY BREAKS

Although trade contractors may resist doing it, placing a capillary break between the footer and the foundation wall is critical to control foundation moisture. Many products can function as the break, including physical barriers like polyethylene sheeting, proprietary products specifically designed for capillary breaks, and spray-applied dampproofing products like elastomeric asphalt or clear sealers. In practice, poly seems to be the most popular capillary break. It is simple to install, inexpensive, and it easily incorporates into the construction schedule.

Backfill

A concrete foundation should be fully cured and all parts of the foundation drainage and moisture proofing installed before backfill begins. Concrete curing time varies with temperature and atmospheric moisture, but usually takes about 10 days.[9] Well-draining, clean, granular material should be installed to within 18" of final grade. To minimize the effects of settling, tamp backfill in 8" lifts to 95% compaction, according to the Standard or Modified Proctor Compaction Test. Atop the backfill, add an 8"–10" layer of low-permeability soil to help convey water from the foundation. Finally, fill the area with topsoil to rough grade and then establish rough grade sloping 5% away from the foundation.

Basement Foundations

Basement foundations are popular in the Northeast and Midwest—areas with substantial heating loads and ground moisture. Although concrete generally can handle intermittent moisture, you want systems to prevent moisture transfer into the home and into moisture-sensitive building materials. Also, because home owners frequently expand their living area by finishing a basement, the basement foundation must be equipped to manage moisture well.

To promote the long-term performance of the foundation system, consider insulating the basement wall during construction even if the code doesn't require basement wall insulation. A well-insulated foundation wall can reduce the energy needed to heat and cool the basement, prevent condensation (and, hence, potential moisture issues), and enhance comfort in the space. Even if the basement will not be used as living space, a solid insulating wall will contribute to whole-house efficiency as ducts and mechanical systems located in the basement face less extreme temperatures.

CONCRETE ADDITIVES

Fly ash, slag cement, and silica fume— by-products of industrial processes—can be repurposed as concrete additives, rather than adding to landfill waste. These concrete additives enhance concrete workability, durability, and strength. Because they also increase curing time, the optimum proportion of additives is seasonally dependent. Larger amounts of additives are desirable in hot, dry weather. Ask your local ready-mix concrete producer which additives are available locally and what the optimum mix is for your performance requirements.

Crawl Spaces

Crawl spaces are an affordable foundation system and a handy location for mechanical systems, ductwork, and storage. Unfortunately, they have a long and dirty history of moisture and mustiness. Nevertheless, you can build crawl spaces that control moisture and enhance energy efficiency, applying the same principles used for basement construction (see Crawl Space Insulation in this chapter). Capillary breaks, a sealed vapor retarder that separates the ground from the space, and good insulation help make unvented crawl spaces durable and energy efficient. If the crawl space will not have a concrete floor, consider upgrading the vapor barrier material to a reinforced product that resists tears better than 6-mil polyethylene does.

Slab Foundations

Slab foundations are common in mild climates and if building codes don't require that they be insulated, they often are not. However, an uninsulated slab wastes energy and provides an opportunity for condensation when its temperature falls below the dew point of interior air. This temperature difference can contribute to mold growth, especially when the floor is carpeted. Moreover, cold slab floors are uncomfortable in almost any climate. Slabs lose heat primarily outward through the perimeter. To avoid these problems, use slab-edge insulation to bring the temperature of the slab closer to a home's interior temperature (see Slab-on-Grade Foundation Insulation in this chapter). Exterior slab edge insulation is typical, but you could instead build a floating slab using interior insulation. As with any foundation system, follow local codes or recommendations for termite- and radon-resistant construction practices.

Closed Crawl Spaces: Introduction for the Southeast (http://www.crawlspaces.org) provides detailed information and construction drawings for building closed crawl spaces in the Southeast. It also includes information about building closed crawl spaces in cold climates.

Insulating Concrete Form Foundations for Basements and Crawl Spaces

An ideal waterproofing system for *insulating concrete forms* (ICFs) uses two levels of protection to prevent water intrusion and to protect the waterproofing system from damage from backfill and tree root growth. Below-grade ICFs require this special attention to waterproofing for three reasons:

1. ICFs' plastic surface is difficult for a waterproofing membrane to adhere to.

2. If leaks occur, they are difficult to find, costly to repair, and often destructive to the foam.

3. If you don't protect them from ground moisture, joints between interlocking foam forms can allow water to enter.

Options for waterproofing ICFs include peel-and-stick membranes and drainable membranes that create an air gap between the foundation and the soil. This gap allows water that may penetrate beyond the membrane to drain to the tile below. Most ICF manufacturers provide a list of compatible materials, guidelines for waterproofing their products, or both, when they are used below grade. The Insulating Concrete Form Association recommends a dual waterproofing system that provides two layers of defense—a protective sealant (such as a self-adhesive product) and a membrane that provides an air gap for drainage between the rigid membrane and the first-line waterproofing material (fig. 3.6).

The secondary waterproofing material that is bonded to the ICF— should water penetrate the first-line dimpled membrane—must be compatible with the foam. Be sure to check the manufacturer's list of compatible materials. Solvent- or petroleum-based spray-applied membranes are generally incompatible with, and can destroy, foam.

Dimpled membranes (typically made from *high-density polyethylene (HDPE)* with post-industrial recycled content) create an air gap between the foundation and the secondary waterproofing membrane. This gap allows water that may penetrate the waterproofing membrane (or condensation from the foundation wall) to drain by gravity and work its way through the foundation drainage system. By alleviating hydrostatic pressure on the foundation, dimpled membranes reduce the likelihood of water being driven into the structure.

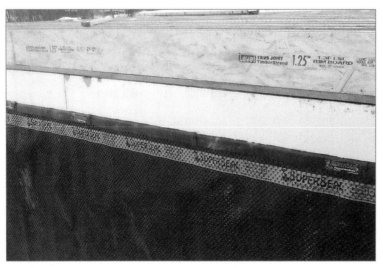

FIGURE 3.6 | ICF foundation with waterproofing membrane

An impermeable HDPE dimpled membrane keeps below-grade water, and the hydrostatic pressure it exerts, away from the foundation. *(Jeannie Leggett Sikora)*

The rigid polyethylene dimpled membranes resist damage, but they can be damaged by construction debris and rocks. Use clean fill and proceed carefully to avoid damage.

Foundation Insulation

Foundation insulation improves home energy efficiency, boosts performance by reducing the potential for condensation at the foundation wall or slab floor, and improves comfort in the home. Builders often overlook foundation insulation because the temperature below grade is relatively constant and, therefore, the foundation is not subject to the temperature extremes experienced above grade. Although insulation needs and methods vary, homes in all climates can benefit from insulated foundations.

Consider below-grade moisture in designing and selecting materials for the foundation insulation system. The insulation system must, first and foremost, resist moisture and be approved for below-grade use. In addition, consider whether the space will be conditioned, if it will contain mechanical systems and ductwork, and, especially for slab foundations, how insulation can make the home more comfortable.

Foundations can be insulated from the interior, exterior, or, as with ICFs, on both sides. When you are building a new home, insulating the foundation from the exterior is an effective and simple solution to foundation insulation. During the heating season, the exterior insulation effectively raises the interior surface temperature of basement, crawl space, and slab foundations to reduce the potential for condensation and resist heat loss.

TERMITES

If you use foam below grade and build in termite-prone areas, you must include a physical barrier to prevent termites from tunneling through the foam to get to the wood material they crave. For areas of moderate to heavy termite activity, the International Residential Code (IRC) contains provisions for foam below grade.[10] It's important that you understand these requirements when using foam below grade.

You can find ideas for finishing exterior foundations at http://tinyurl.com/insulationhowto.

Basement Insulation

When insulating basements from the interior, use materials that can handle some moisture. The insulating system should be designed to dry out if it becomes wet. In the heating season, it should prevent the relatively warm moist basement air from condensing on the relatively cool basement walls. Because materials in basement insulation systems must frequently dry to the interior, researchers recommended that the insulation be at least *semi-vapor-permeable* (have a vapor permeance greater than 0.1). The wall system's ability to dry increases with increasing permeability. A higher perm rating indicates better ability to dry if it becomes wet. According to Building Science Corporation, materials that can be used to achieve a semi-vapor-permeable insulation system include:[11]

- Up to 2" of unfaced, XPS foam (perm about 0.5)
- Up to 4" of EPS foam
- Up to 3" of closed-cell spray foam
- Up to 10" of open-cell spray foam (e.g., Icynene)

You can design the wall system to endure occasional moisture and maintain its durability and energy efficiency.

Basement Insulation Systems (http://tinyurl.com/InsulatingBasements) includes details for several basement insulation systems that can withstand the unique basement environment. Climate-specific design details can be found in EEBA's *Builder's Guide* series.

Crawl Space Insulation

Most crawl spaces are vented, have earthen floors, and frequently house ductwork, piping, and mechanical equipment. They are prevalent in the Southeast and Midwest where natural moisture abounds. Using traditional construction methods for these spaces promotes dampness as water vapor migrates through the earthen floor and humid ventilation air infiltrates during the summer. Further, cold ventilation air in the winter causes excessive energy loss from ducts and mechanical systems, and other trouble, like frozen pipes. Research has shown that crawl spaces perform best—from both a moisture and energy standpoint—when they are built like short basements. By thermally isolating the crawl space from the exterior and including moisture-, pest-, and radon-resistance measures, a crawl space can be clean, dry, and an integral part of a high performance home.

Slab-on-Grade Foundation Insulation

Energy engineers often refer to concrete as a "heat sink" because of its natural ability to absorb and store thermal energy. Because thermal energy travels from warmer to colder areas, a massive concrete slab can suck heat away from the inside of a home. In mild and cooling-dominated climates, this connection with the ground may not cause discomfort or high energy bills; hence, the 2012 International Energy Conservation Code[12] does not require slab-edge insulation in climate zones 1–3. Insulate the slab to IECC levels or better to cut energy bills, improve comfort, and reduce condensation potential (and the accompanying risk for moldy carpet or other moisture-related problems that can occur in finished basements). Figure 3.7 shows a diagram of slab-edge insulation.

Most heat loss from a slab on grade occurs outward from the edge of the slab. Therefore, at a minimum, insulate the slab edge with rigid insulation, either on the exterior or interior (in a floating slab installation) of the footing, to improve building durability and energy efficiency. DOE's *Slab Insulation* fact sheet (http://www.nrel.gov/docs/fy01osti/29237.pdf) describes and depicts interior and exterior slab-edge insulation methods. Full horizontal insulation coverage under a slab can improve comfort even more, and cut energy bills and reduce the potential for condensation. Building codes require

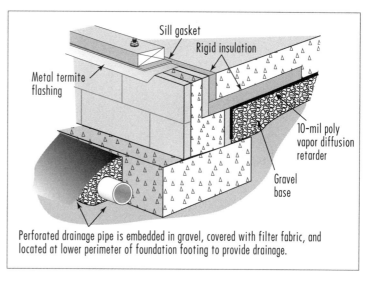

Sill gasket

Rigid insulation

Metal termite flashing

10-mil poly vapor diffusion retarder

Gravel base

Perforated drainage pipe is embedded in gravel, covered with filter fabric, and located at lower perimeter of foundation footing to provide drainage.

FIGURE 3.7 | Slab-on-grade insulation

This figure illustrates the floating slab method for insulating a slab edge and protecting the foundation from moisture migration. *(U.S. DOE)*

full under-slab insulation in all climate zones if radiant heating tubes are run through the slab.

You can use proprietary products to insulate the slab edge. Foam insulation in these products is likely to be treated for termite-resistance, which can help meet local code requirements for below-grade foam.

Precast Concrete Foundations

Precast concrete foundation walls (fig. 3.8) can create energy-efficient basement and crawl space walls. Because precast foundations are resource and energy efficient and they install quickly (minimizing soil exposure), they can earn points in several categories within green certification programs.

FIGURE 3.8 | Precast concrete basement foundation

Precast concrete foundations earn points in several categories within green certification programs. *(Jeannie Leggett Sikora)*

Although the high-strength (5,000 *psi*) concrete used in precast walls resists moisture better than site-poured concrete, Section R406 of the 2012 IRC still requires dampproofing or waterproofing. Check the manufacturer's construction details for applying foundation waterproofing correctly.

Frost-Protected Shallow Foundations

Frost-protected shallow foundations (FPSFs), in which footers do not extend below the frost line, are practical and energy efficient. Essentially, FPSFs use vertical and horizontal insulation to keep heat from dissipating into the ground, which keeps the footers warmer (fig. 3.9). They essentially raise the frost line to allow shallower footings. Compared with conventional stem walls, FPSFs minimize disturbance to the natural environment, reduce excavation costs, lower the amount of material used in foundation construction, and improve energy efficiency. Section R403.3 of the 2012 IRC discusses frost-protected shallow foundation design.

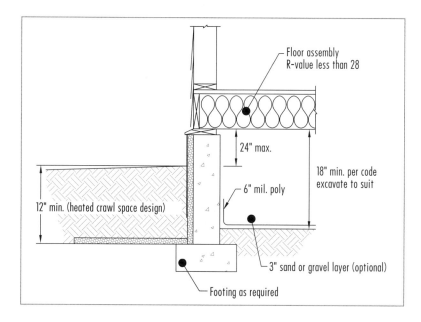

FIGURE 3.9 | Frost-protected shallow foundation

Sample detail for a frost-protected shallow foundation. *(NAHB Research Center)*

Floor assembly R-value less than 28

24" max.

18" min. per code excavate to suit

6" mil. poly

12" min. (heated crawl space design)

3" sand or gravel layer (optional)

Footing as required

FRAMING: STREAMLINING THE STRUCTURE

Many green homes are "stick framed," just as most conventional homes are. But green home structures differ from conventional homes because they often

- use less framing material;
- allow more space for insulation;
- provide for mechanical distribution systems in conditioned space;
- incorporate certified lumber from sustainable forests; and
- use engineered wood manufactured from young, fast-growing trees which is made with binders that don't compromise IAQ.

Advanced Framing Methods

Many homes are "over framed," which wastes lumber. How many times have you seen three studs in a row next to a window, or back-to-back studs in a clear wall section? Whether because of a mindset, ("More lumber makes the building stronger, so it must be better"), a lack of planning, or for convenience, many walls contain more lumber than is necessary. Using lumber beyond what the structure requires eats up space for insulation.

Panelized wall systems and modular homes help reduce lumber and jobsite waste because they are built in a strictly controlled factory environment.

Advanced framing, or *Optimum Value Engineering (OVE)*, minimizes the amount of framing lumber without compromising structural performance.

Eliminating unnecessary lumber from the exterior structure creates more room for insulation, which creates a better insulated *building envelope.* There are many methods to accomplish OVE. These include *ladder blocking* where interior walls adjoin the exterior walls, and using *header hangers* instead of *jack studs* to support window headers. You can incorporate methods as needed; advanced framing is not an all-or-nothing proposition.

Use these advanced framing methods as appropriate and as allowed within building codes to reduce lumber usage and improve energy efficiency:

- Design walls and roofs in 2' increments to incorporate commonly sized construction materials (e.g., 4 × 8 sheets of OSB and drywall, 8' studs) to reduce waste.

- Space studs at 19.2" or 24" o.c.

- Use two-stud corners (with scrap lumber nailers or special clips to attach drywall).

- Use in-line or "stack" framing to align wall studs vertically with members directly below (fig. 3.10). Because loads transfer directly downward, this practice eliminates the need for a double top plate.

FIGURE 3.10 | In-line framing

Lining up floor joists with framing members spaced 24" on center is one of many advanced framing techniques that save lumber and increase a wall's overall R-value. *(Jeannie Leggett Sikora)*

- Eliminate headers in non-load-bearing walls.
- Add a layer of insulation to headers (in lieu of an additional framing member).

Builders have used advanced framing techniques for decades, but as with any construction practice you are unfamiliar with or have not used, check code requirements before implementing.

Housing Ductwork

Best practice for high performance homes dictates running all ducts and mechanical equipment in conditioned space with no distribution piping or ducts running through exterior wall cavities. To successfully incorporate ducts in a home's conditioned space, consider mechanical system layout and duct design before finalizing the framing plan. Some builders use open-web floor trusses (fig. 3.11) and drop ceilings (fig. 3.12) to facilitate placing ducts in conditioned space. This requirement is especially relevant when you are using engineered lumber; it can't always be notched or cut to accommodate mechanical systems.

Advanced Wall Framing (http://tinyurl.com/ AdvancedFraming) is a fact sheet from DOE with diagrams that illustrate advanced framing strategies.

Sustainably Harvested Lumber

Originally intended to stem the tide of tropical deforestation, forest certification programs have caught on in North America as a way to ensure sustainable forest management. The United States has three distinct forest certification programs: The America Tree Farm System, the Sustainable Forestry Initiative, and the Forest Stewardship Council (FSC). Although each program is different, all have the key elements to be credible (e.g., third party verification) and they improve forest management, according to a 2008 policy statement by the National Association of State Foresters. Certified wood is widely available at home improvement stores and many local lumberyards.

One of the design challenges for green homes is bringing ducts into conditioned space. California's *Home Builders' Guide to Ducts in Conditioned Space* (http://www.energy.ca.gov) provides clear guidance, diagrams, and information.

FIGURE 3.11 | Open-web floor truss

Open-web trusses facilitate running ducts in conditioned space between floors. To keep any duct air leakage within the building, carefully seal and insulate the rim joist area. *(Jeannie Leggett Sikora)*

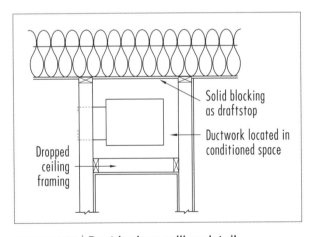

Solid blocking as draftstop

Ductwork located in conditioned space

Dropped ceiling framing

FIGURE 3.12 | Duct in drop ceiling detail

One method for keeping ducts in conditioned space is to create a drop ceiling with an air barrier and insulation separating the duct chase from unconditioned space. *(U.S. DOE)*

Engineered Wood Products

Because they are a resource-efficient natural building material, and because renewable energy is used in their manufacture, many engineered wood products earn points in green building certification programs. Most of the leading engineered wood companies have readily available literature that lists points for which their products qualify under each major green building program.

Engineered wood products are a green building alternative to dimensional lumber. Engineered wood is consistently sized, *dimensionally stable*, and resource efficient. It includes softwood and hardwood from fast-growing "scrub" trees instead of older-growth trees, reclaimed wood waste (e.g., sawdust, scrap wood), or crop fibers (e.g., wheat straw).

Engineered wood has drawbacks, however. One is cost: engineered wood products typically cost more than dimensional lumber. The other is the binders: they contain formaldehyde, a *volatile organic compound (VOC)* that has been linked to negative human health effects. However, the concern over VOCs may be overstated; engineered wood structural materials contain low-emitting phenol-formaldehyde binders rather than the more volatile urea-formaldehyde binders historically found in furniture and cabinetry. Because of these low emissions, most meet or are exempt from formaldehyde emissions regulations. For example, plywood and OSB built to U.S. Voluntary Product Standards PS1 and PS2 easily meet the most stringent international formaldehyde emissions standards.[13] Other engineered wood products are exempt from new U.S. legislation covering formaldehyde emissions. The Formaldehyde Standards for Composite Wood Products Act exempts structural panels, structural composite lumber, engineered I-joists, and glued laminated lumber from its regulations.

The NAHB Research Center maintains a growing database of green products (http://www.greenapprovedproducts.com) that identifies the section and number of points within the NGBS for which each approved product is eligible.

Engineered wood building products include

- structural plywood;
- oriented strand board (OSB);
- I-joists;
- laminated strand lumber (e.g., rim board);
- laminated veneer lumber—studs, beams, headers;
- finger-jointed studs;
- roof and floor trusses;
- engineered wood flooring; and
- engineered wood siding.

Formaldehyde and its effect on human health have long been studied. Most engineered wood products do not use highly emitting urea-formaldehyde binders. Also, because urea-formaldehyde is water soluble, it is not incorporated in products designed for exterior use. Commonly used *phenol-formaldehyde* and *methylene diphenyl diisocyanate (MDI)* binders emit fewer VOCs than do urea-formaldehyde binders. When you are not sure what type of binder was used, check the material safety data sheet (MSDS) or specify exterior-grade engineered wood.

APA GREEN VERIFICATION REPORTS

APA-The Engineered Wood Association has begun issuing Green Verification Reports to provide information about how engineered wood products fare under the green building rating systems. Much like an International Code Council Evaluation Service (ICC-ES) Evaluation Report examines how a product comports with existing standards, APA's Green Verification Reports identify points a product qualifies for under national green rating programs such as NGBS and LEED for Homes.

Wood Framing Alternatives

Alternatives to conventional framing—like *structural insulated panels (SIPs)* or ICFs— can reduce material usage, increase insulation, simplify air sealing, and reduce construction time and labor use. Many builders have successfully incorporated these systems into their green homes. Although they may encounter a learning curve, some builders find their niche using alternatives to framing lumber and reap benefits such as fast assembly. The decision to try an alternative to conventional framing depends on a host of factors such as local cost for labor and materials, climate, and willingness to try something new.

Steel Framing

Steel is strong, straight, dimensionally consistent, termite-resistant, recyclable, and durable. However, it is also comparatively expensive, conducts heat readily, and can be a problematic alternative for conventional wood framers not accustomed to using it. Design and educational materials and new product developments can help overcome the latter two issues. Prescriptive tables for building homes with steel framing are in the IRC. The Steel Framing Alliance (http://www.steelframing.org) has published design manuals, including *Thermal Design and Code Compliance for Cold-Formed Steel Walls*[14] to help builders create highly energy efficient steel-framed walls.

The Foam Sheathing Coalition offers a fact sheet, *Guide to Attaching Exterior Wall Coverings through Foam Sheathing to Wood or Steel Framing* (http://tinyurl.com/FoamSheathing) with prescriptive information for siding attachment over foam sheathing (to steel or wood framing). Some manufacturers have developed steel insulated panels (similar to SIPs) and steel-framed wall panels with integrated foam insulation.

Structural Insulated Panels

SIPs are wall panels that look like OSB sandwiches—with two layers of OSB encasing several inches of a solid foam core. Some products come prefinished with drywall or tongue-in-groove interior paneling. SIPs, although they require careful detailing to avoid air infiltration, produce very little on-site waste and provide a continuously high R-value wall (and ceiling if also used for roof panels). The foam frequently contains recycled content; check with individual manufacturers for details.

Although SIPs are more common in cold climates, they have gained popularity in other climates as well. Builders use SIPs for roofs in unvented attic designs, allowing them to place ducts into a conditioned attic space (fig. 3.13). Although the panels themselves are an air barrier, connection sealing details are the crux of creating a tightly constructed SIPs home. The Structural Insulated Panel Association (http://www.sips.org) offers CAD drawings of SIPs connections you can use in construction documents.

FIGURE 3.13 | SIPs roof

SIPs in a roofing application enable this builder to easily place ducts in conditioned space. *(Scott Homes)*

The IRC includes prescriptive guidelines for SIP walls.

Insulating Concrete Forms

Just as they are a viable option for foundations, ICFs also create energy-efficient, strong, and durable above-grade walls. Where applicable, ICFs are eligible for points towards green building certification for their

- energy efficiency;
- recycled content plastic ties that join the forms;
- recycled content (fly ash) concrete;
- recycled content reinforcing steel;

- indigenous materials (typically if concrete is manufactured and sourced regionally);
- resource efficiency (using fewer materials than a poured-concrete wall and producing little on-site waste);
- mold- and moisture-resistant building materials; and
- materials that emit no formaldehyde.

ROOFS: TOPPING IT OFF

The primary role of a roof is to keep out rainwater. Yet a roof can also play a pivotal role in a home's environmental performance. Roofs on green homes can enhance durability and energy efficiency (such as raised-heel trusses that permit full-height insulation at the eaves), and can include systems for capturing and reusing incident rainwater. Roofs also are a prime location for solar panels. One type of roofing material, with *building-integrated photovoltaic (BIPV)* panels, can both protect a home from the weather and use the sun's energy to produce electricity. Finally, vegetated roofs and recycled-content roofing materials are becoming more common and may be worth investigating for a green home project.

Roofing materials have come a long way, with newer asphalt shingles offering warranties of 50 years or more. Yet while material durability is an essential element of a long-lasting roof, installation can be the more critical factor in longevity.

Durability is in the Details

Flashing is important for defending against water intrusion. Because flashing protects the wood structural materials such as roof and wall sheathing, APA-The Engineered Wood Association has a library of free flashing construction details (http://www.buildabetterhome.org).

For a green home, roof flashing details should be called out on the plans. Trade contractor specifications should include these details as the first step to ensure proper installation. Include flashing details for the rakes and eaves, roof-to-wall intersections (including kickout flashing when the wall abuts a gutter), valley areas, penetrations, chimney-to-roof, and dormer flashing.

Drip edge flashing pushes water away from the roofline, prevents wind-driven rain from getting under the shingles, and helps keep rainwater off the fascia board. Because it protects the roof's structural elements from moisture damage, drip edge is vital to the roof system's durability.

For moisture protection, drip edge flashing should be installed at all eaves and rakes. Outside of high-wind areas, it should be nailed directly to the roof sheathing along the eaves (with roofing underlayment or ice guard extending over the metal flashing) and on top of the underlayment on the rakes. In high-wind areas, building codes may require attaching drip edge on top of the underlayment with nails and roofing cement.

In snowy climates with freezing temperatures, ice dams—and their associated roof damage and water leaks—are not uncommon. Carefully designed energy-efficient homes typically prevent icicles from forming by preventing heated air from leaking into the attic and by ensuring adequate attic ventilation to keep the attic cold in winter. Nevertheless, a thick membrane at the eaves will provide further insurance against damage from ice dams and wind-driven rain. Typically, these membranes are self-adhesive, *bituminous*, and resist tears and slipping. Most codes require ice dam membrane to extend 2' up the roof beyond the exterior wall line. Waterproofing membrane materials can also provide an extra layer of protection anywhere there will be flashing, such as at roof valleys and ridges.

INCORPORATING DETAILS ON PLANS

Roof details and specifications vary by climate, roof slope, and roofing material. Many roofing manufacturers have stock details that you can easily incorporate into a set of plans. Roof flashing and other details that you can incorporate into your plans are at http://tinyurl.com/stockdetails. Although most details are manufacturer or product specific, you can adapt them for your application. The National Roofing Contractor Association (NRCA) offers details specific to asphalt shingles at http://tinyurl.com/AsphaltRoof.

Because many ice and water membrane products are not vapor permeable, design the roof and attic system so any moisture that accumulates in the sheathing can dry (provide adequate attic ventilation in a vented roof design, for example). This is especially important if the entire roof will be covered.

A growing variety of specialty products for roofing underlayment for specific applications, in addition to the widely used 15# roofing felt, has emerged. Check with manufacturers for products specifically formulated for high temperatures (e.g., the desert Southwest), low-slope roofs, unvented attics, or metal roofing.

Moisture-Resistant Roof Sheathing

Roofing alternatives that combine moisture resistance and sheathing in a single product are gaining popularity. They can eliminate the need for a separate underlayment, except in cases where the building code requires ice and water membrane.

Energy Efficiency

Although the roof's main function is protecting the house from rain, a roof is also part of the building's energy management strategy. Roof overhangs contribute to the durability of the walls and can be a passive solar strategy (e.g., shading south-facing windows in summer but allowing solar heat gain in winter).

Roofing design elements that impact energy efficiency include attic ventilation, framing member design, overhangs, and specific materials.

Cool Roofs

Dark-colored roofs, just like any dark surface, warm up when the sun shines on them. Light-colored or specialty coated roofing materials absorb less solar energy. Lower roof temperature lowers heat gain in the attic and reduces cooling energy used. Savings depend on factors such as climate, whether ducts and mechanical equipment are located in the attic, and the amount of duct and attic insulation. For example, although California's Title 24 building code mandates the use of cool roofing materials for new residential construction, it exempts homes without ducts in the attic, those with adequate levels of attic insulation, and homes with radiant barriers installed in the attic, because the energy savings don't justify the expense.

Cool roofing materials function as they do because of two thermal properties: *solar reflectance* and *thermal emittance*. Materials with high solar reflectance do not absorb as much solar energy as conventional materials do. Materials with high thermal emittance can cool rapidly because of radiant losses to the atmosphere (e.g., when the sky is clear at night). In cooler climates, cool roofing materials may extract a *heating penalty*. However, if you do not place ducts or mechanical equipment in the attic and you build a well-ventilated pitched attic, the heating penalty from cooler roof temperatures will be small. Because cool roofing materials lower roof surface temperatures (creating condensation potential), use climate-specific design strategies to ensure the roof system can handle moisture.

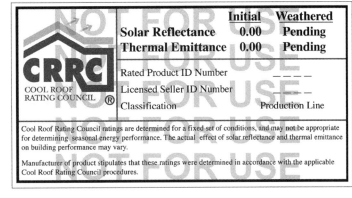

FIGURE 3.14 | CRRC label

The Cool Roof Rating Council administers a third-party testing and labeling program to identify the thermal properties of roofing products. *(Cool Roof Rating Council)*

With recent technological advances, there are cool roofing options in almost any color and type of roofing material. For any application, the easiest way to find cool roofing products is to look for the ENERGY STAR label. Low-slope products (≤2:12) that meet ENERGY STAR's current criteria have a solar reflectivity of 0.65 or greater (0.5 after three years), and steep slope (>2:12) materials have a reflectivity of ≥0.25 (0.15 after three years). For more detailed information, such as the product's thermal emittance, look for the Cool Roof Rating Council label (fig. 3.14) or search the council's online directory of rated products (http://www.coolroofs.org). The council does not set criteria for a cool roof; it provides objective information about roof materials' thermal properties. The council oversees the testing and labeling of roofing products, sets standards for testing, and accredits third-party testing laboratories.

For an overview of cool roofs, see DOE's *Guidelines for Selecting Cool Roofs* (http://www1.eere.energy.gov/femp/pdfs/coolroofguide.pdf).

Climate plays an important role in the moisture and thermal performance of cool roofing systems. An architect or building scientist

You can estimate the energy savings from implementing cool roofing technology using an online calculator (http://www.roofcalc.com) that U.S. DOE has developed.

can create or review the designs for using cool roofing materials. If you are considering using cool roofs in temperate or cold climates, have your designer model the *hygrothermal behavior* of the assembly (e.g., using the WUFI software program[15]) before finalizing the design.[16]

Radiant Barriers

Radiant barriers can reduce heat gain in the attic. They are most effective in homes in hot climates where ducts and mechanical systems are located in the attic. In this case, they can save about $150–$200 in annual cooling costs. Savings dwindle as the climate moderates and in homes with no HVAC equipment in the attic.[17] A radiant barrier looks like aluminum foil or a space blanket. Radiant barriers are installed at the roofline and may be laminated to the underside of roof decking as an all-in-one sheathing and radiant barrier product.

Although they are both used to lower attic temperatures, radiant barriers and cool roofing materials function somewhat differently. Whereas cool roofing reflects solar radiation and has a high thermal emittance to radiate the energy outward, radiant barriers have low thermal emittance (<0.1). They effectively prevent solar energy that the roofing system absorbs from emitting into the attic. For a radiant barrier to prevent this thermal emittance and function as designed, there must be an air space between the reflective surface and the insulation below the barrier. Without this air space, the thermal energy can simply conduct to the material adjacent to it.

Some product manufacturers recommend installing the radiant barrier atop attic floor insulation with the shiny side up. While the material is new and shiny, the application can be effective at keeping heat inside the house in winter (because it thwarts thermal emittance toward the cooler attic). However, as the barrier accumulates dust, its thermal emittance increases. Models indicate the material's effectiveness decreases by 50% in 1 to 10 years.[18] Further, if the radiant barrier is not vapor permeable, a horizontal application can also trap moisture in the attic insulation.

Full-Height Attic Insulation

Insulation works primarily by trapping motionless air, which is a poor conductor of thermal energy. A common problem with sloped roofs is that insulation compresses at the eaves so it does not retain its pockets of still air and, hence, its rated R-value. To create sufficient space for full-height insulation at the eaves, builders can use *raised-heel trusses* (also called energy heel trusses) or cantilever conventional trusses over the edges to create extra space for insulation above the exterior wall (fig. 3.15). The cantilever method also creates overhangs that enhance durability by protecting the walls below from rainfall and *UV radiation*.

FIGURE 3.15 | **Cantilevered truss**

Using "oversized" roof trusses that cantilever over an exterior wall creates additional space for insulation and adds overhangs that help improve the wall's durability. *(Jeannie Leggett Sikora)*

Unvented Attic Design

In recent years, unvented attic designs have garnered attention as a solution for bringing HVAC equipment and ducts into conditioned space, preventing air infiltration between the top-floor living space and the outdoors, and minimizing the risk to structures from wildfires. In these designs, builders install insulation at the roofline and the attic becomes semiconditioned space.

Although it took years of debate to quell concerns about the need for attic ventilation, unvented attic designs are now allowed by code. In fact, building designers often recommend unvented attics for hot and humid climates where bringing warm, moist ventilation air into attics that often contain ducts and air handlers defies common sense. The National Roofing Contractors Association says unvented attic designs are a viable alternative throughout the United States and urges that designers consider using them in hot, humid climates.

Shingle manufacturers initially were wary of unvented attics and often negated their products' warranties for unvented designs. Now, however, most offer full warranties for their products when shingles are installed according to code requirements and manufacturer's instructions on an unvented attic design. Still, check the manufacturer's warranty requirements for unvented attics.

Unvented attics must be carefully constructed so the attic is "connected" to the conditioned space of the house rather than to the outdoors. The attic should be within the home's thermal (insulation) and pressure (separation of outside air from inside air) boundaries.

Researchers at the University of Waterloo and Building Science Corporation used computer models to evaluate various insulation and air-sealing strategies for unvented attic designs and to establish climate-specific boundaries for these designs. Their report, *Moisture-Safe Unvented Roof Systems*, can inform decisions about unvented roofs for your climate.[19]

Unvented attics produce the most energy savings in homes in which ductwork and air handlers are located in the attic, the ducts are relatively leaky, and in cooling-dominated climates. Isolating the impact of unvented attics on whole-house energy savings is difficult, according to a study by the National Renewable Energy Lab that identified numerous other influencing factors such as climate, attic insulation value, air infiltration, and reflective value of the roofing material.[20]

Finally, simply locating ducts in conditioned space does not eliminate the need for building a tight duct system. Both comfort and durability require that the air travels where it needs to; leaky ducts can thwart that effort.

Solar Energy Generation

Even if solar panels are not part of the installation in a new green home, consider roughing in the electrical and plumbing lines so solar panels can be installed in the future. This extra effort during construction can greatly reduce the expense and disruption of any future roof-mounted renewable energy systems. Chapter 8 addresses on-site renewable energy systems.

Vegetated Roofs

Commonly called green roofs, vegetated rooftops are, literally, sprouting up all over the place. This type of roof has been popular in Europe for decades and green roofs are increasingly common in the United States. Although they are more prevalent on commercial buildings, green roofs are appearing on homes. They are not limited to flat roofs.

Vegetated roofs offer several benefits, including thermal insulation, heat mitigation (the plants provide evaporative cooling of the roof), storm water management, and protection of the waterproofing membrane from UV radiation damage.

However, because the systems add tremendous weight to a roof, a structural engineer must approve the design. Although ANSI standards address vegetated roofs, they apply to fire and wind uplift; there is no prescriptive design method for vegetated roofs.

Going Green with Other Roofing Materials

You don't need a green thumb to build a green rooftop. You can simply select roofing products for their green properties. For example, you can use locally sourced natural materials like wood shakes or natural slate. Of course, these materials generally cost more than asphalt shingles. Metal roofing, another premium product, is manufactured using recycled content (the percentage of pre- and postconsumer waste varies by type of metal and manufacturer). It's durable and recyclable. Other recycled-content roofing materials are available and the array of options continues to grow. Trade shows and green building suppliers are good resources for keeping up with the options.

Asphalt shingles also have come a long way in durability, recycled content, and recyclability. Several states—concentrated in the Midwest and Southeast—offer shingle recycling from roof tear offs. Some states also have programs for postindustrial asphalt shingle recycling. To find out about shingle recycling in your state, click the interactive map at http://www.shinglerecycling.org. There is a large potential market for recycling shingles into hot-mix asphalt for roadways and other materials.

The asphalt shingle recycling market will probably continue to grow in North America because it makes economic sense, according to William Turley, executive director of the Construction Materials Recycling Association.[21] According to Turley, hot-mix asphalt producers save about $2–$6 per ton by incorporating recycled shingle content into their products.

WINDOWS

Ample windows, with their views and natural light, are a desirable feature. However, the large expanses of windows popular in new homes add to heating and cooling costs. Although residential window technology has improved, windows have a lower insulating value than the surrounding wall. The best triple-pane double-hung windows on the market have an R-value of about 7.[22] Typical double-pane windows have an R value of 3. Compare this with an R-value of at least 13 for the surrounding wall. Except for thoroughly passive solar design, in which the windows serve as solar energy collectors, every window is a missed opportunity to increase a home's energy efficiency. Of course, aesthetics as well as efficiency influence green home design.

You can improve energy efficiency by selecting products with the ENERGY STAR label and by using the information on the National Fenestration Rating Council (NFRC) sticker included on most windows. ENERGY STAR applies climate-specific efficiency criteria to windows so you will be making an energy-efficient choice for your region (fig. 3.16). The NFRC label includes some or all of the following information:

U-Factor

U-factor describes the ability to conduct heat. A lower U-factor means a more energy efficient window. U-factor is inverse to R-value, which describes resistance to heat flow (as with insulating material). A U-factor of 0.33 is equivalent to an R-value of 3. The best triple-pane window on the market has a U-factor of about 0.09 (R-11). U-factors apply to an entire window unit so be wary of manufacturer's statements about center-of-glass insulating values, which do not represent a window's installed performance.

Solar Heat Gain Coefficient

SHGC is a number between 0 and 1 that represents the portion of solar radiation that can be transmitted indoors as heat. The ideal number depends on climate and whether or not a structure uses passive solar heating. For homes in cold climates, higher SHGC will allow more solar heat gain. Lower SHGC windows are desirable in warmer climates. Passive solar homes employ high SHGC windows (e.g., 0.6 or higher) on the southern exposure to collect as much solar energy as possible when the sun is low in winter, and rely on shading devices to block solar gains in summer.

Visible Transmittance

Visible transmittance (VT) describes how much daylight enters a home. Specialty glass coatings that enhance a window's thermal properties typically reduce VT, which is a number between 0 and 1. Typical values range from about 0.35 to 0.70 depending on the number of panes and the coatings used (e.g., low-e, tinting). Spectrally selective low-e coatings, those that enable part of, but not all, UV spectrum through, have VT values that rival those of clear glass.

Air Leakage

Although it is not reported as commonly as the three other attributes, air leakage is a useful indicator of a window's installed performance. The optional air leakage (AL) rating indicates the amount of air that will seep through cracks in the window assembly—with higher AL ratings indicating a leakier window. The Efficient Window Collaborative (http://www.efficientwindows.org) recommends looking for windows with an AL of 0.30 cfm per square foot or lower.

Condensation Resistance

Condensation resistance (CR), like air leakage, is reported to a lesser extent than U-factor, SHGC, and VT. CR is a number between 0 and 100 (generally between 30 and 80) that represents how well the window resists the formation of condensation (which is directly related to U-factor and air leakage). A higher number indicates more resistance, which can improve energy efficiency and product durability.

FIGURE 3.16 | Sample NFRC label

The NFRC label reveals details about a window's performance. *(National Fenestration Rating Council)*

Installation

How windows and doors are installed significantly impacts their energy efficiency and durability. Although window manufacturers have improved products to address energy efficiency and water penetration issues, most windows can leak. Therefore, you need to prepare rough openings by flashing and sealing them properly to handle any water that gets between the window and the rough opening (fig. 3.17).

Window flashing products abound. Installation, particularly with flexible peel-and-stick sill flashing materials, has become easier. Although the IRC specifies

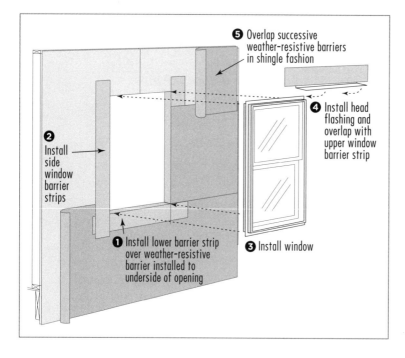

FIGURE 3.17 | Window installation flashing

Most flashing and house wrap manufacturers have readily available window flashing details you can easily incorporate into a set of plans. *(U.S. DOE)*

that windows be installed according to manufacturer's instructions, pay attention to the house wrap and flashing tape manufacturers' instructions as well. When designing window flashing systems, you must consider what will happen to water that gets behind the siding or seeps under windows and create a path to keep it away from the wood sheathing and framing. Also, integrate the flashing tape with the house wrap so water is diverted from the window or door opening. Many flashing and house wrap manufacturers have readily available CAD details you can pull into your architectural drawings. Including these details communicates proper practices to trade contractors and earns credit toward a green building rating.

You can search products by U-factor, SHGC, and VT on the NFRC Certified Products Directory (http://www.nfrc.org/getratings.aspx).

BUILDING "TIGHT"

Homes must be constructed tightly to be energy efficient. Thanks to years of practice, air-sealing product developments, stricter codes and standards, and educational efforts, new homes are being built tighter; in fact, many production homes today are built more tightly than would have seemed possible just a few years ago.

"Build tight and ventilate right" is the mantra for building energy efficient homes for good reason. Unplanned air infiltration, which was long considered a given in home construction, actually ignores sound building science:

- It provides inconsistent ventilation levels—the temperature difference between indoor and outdoor air drives the leakage, which means most leakage occurs simultaneously with the greatest heating- and cooling-load periods.

- Leakage can draw air through dirty and dusty spaces (e.g., insulated wall cavities).

- Leaking air contains water vapor, which can condense into liquid water and cause moisture problems (fig. 3.18).

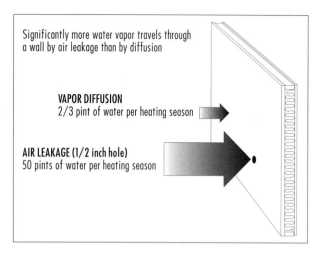

Significantly more water vapor travels through a wall by air leakage than by diffusion

VAPOR DIFFUSION
2/3 pint of water per heating season

AIR LEAKAGE (1/2 inch hole)
50 pints of water per heating season

FIGURE 3.18 | Moisture migration through air leakage

Compared with vapor diffusion, air leakage moves water vapor much more efficiently. *(U.S. DOE)*

Although conventional wisdom promoted allowing buildings to "breathe," modern building science has proven that natural air leakage is unreliable, can compromise building durability, and increases energy consumption. Instead of allowing homes to breathe, green home builders prevent air infiltration through the use of air barriers (solid materials that prevent air movement between unconditioned and conditioned space) and control fresh air exchange via mechanical ventilation. This tight construction and consistent, controlled, fresh air ventilation (and removal of pollutants at their source through spot ventilation) is essential for any green building.

Creating an air barrier in the plane of the exterior wall reduces both whole-house air leakage and *wind washing* of fibrous insulation that compromises insulation's R-value. (Air movement compromises the ability of insulation to maintain dead air space.) Creating an air barrier on the home's interior, especially in heating-dominated climates, keeps warm, moist air inside the building from entering the wall cavity or attic, where it can condense and create moisture issues.

You can incorporate air barriers at both the exterior and interior walls, but be careful that both do not retard vapor diffusion. You need to consider the entire wall assembly, the climate, and the HVAC system when designing a wall system that can dry (e.g., Will absorbent cladding on the exterior create a vapor drive? Does the interior assembly permit vapor to flow inward, where the cooling system can handle drying of the latent load?). You do not want any water vapor that gets into the wall cavity to be trapped there. You can follow prescriptive code recommendations or have a professional design the wall assembly to encourage any entrained water vapor to diffuse outward through vapor-permeable materials.[23]

Installing continuous air barriers on both sides of a wall can be challenging for a few areas of the home, such as where porch framing abuts a wall, behind tubs or showers installed before drywall, or at exterior walls adjacent to stairways. Although these areas require installing an air barrier at different but specific points in the construction schedule, many builders have developed practical solutions for getting the job done. They may use materials that are readily available when the air barrier is constructed or assign non-traditional responsibilities to trade contractors to ensure the process is completed.

Exterior Air Barriers—House Wrap and Solid Sheathing

House wrap, solid sheathing products, and other exterior sheathings (e.g., rigid foam and insulated structural sheathing products) that are taped at the seams with manufacturer-recommended tape are a good starting point to create an exterior air barrier for wood-framed walls. Exterior air barriers should continue all the way around the house. They should be installed prior to windows and before installing any framing that abuts the exterior wall (e.g., porches, roofs). Exterior air barriers that also function as protection from water intrusion must cover the entire building, including the exterior of non-conditioned spaces such as the gable ends and garages.

Air barriers must be installed properly to perform as intended. When using house wrap as an air barrier, its vapor permeability must be adequate for your climate so that if the wall system gets wet, the water vapor can diffuse out of the wall system (including out of unvented cladding if it exists). For very cold climates where an interior vapor barrier is used, the house wrap should be vapor permeable. Greater than 10 perms is considered vapor permeable, but a higher perm rating will allow more vapor to escape.[24] Recognize, however, that water vapor can flow through vapor-permeable materials in either direction. Therefore, the structure

The 2012 International Energy Conservation Code (IECC) and IRC, which have begun to refer to air barriers and thermal barriers, specify whole-house air leakage limits. The 2012 IRC mandates air tightness levels of 3 *ACH50* for climate zones 3 and up. Many builders will have to ramp up their air-sealing measures.

requires an adequate air space between the house wrap and absorbent claddings like brick, since water vapor can pass through the house wrap in either direction. As with any wall system, consult the codes or a design professional for climate-specific designs that promote drying. In general, for house wrap to serve as an air barrier (rather than just preventing water intrusion), follow these practices:

- Wrap it tightly around the entire building and cover all window and door openings, utility penetrations, interior and exterior corners, and joist areas (fig. 3.19).

- Extend the house wrap below the sill plate over the foundation at least 2" and overlap it horizontally (shingle style) and vertically, according to the manufacturer's instructions (typically 6" vertical and 12" horizontal overlap).

- Tape all seams with manufacturer-approved tape.

Do not confuse air barriers, which prevent the movement of air, with vapor barriers, which inhibit the movement of water vapor.

Where the house wrap overlaps the sill plate, you can get the best air infiltration protection by using a manufacturer-approved adhesive to seal the bottom of the house wrap to the foundation. Especially when house wrap is sealed at the bottom, it is critical that you flash window, door, and utility penetrations correctly to ensure water drains on the exterior of the house wrap. Roof-wall intersections may require other details.

FIGURE 3.19 | House wrap installation

House wrap requires certain installation details, such as overlapping at corners and the bottom plate, to form an exterior air barrier. *(Photo courtesy of Treasure Homes)*

Exterior Air Barriers—Sealing at Exterior Wall

To complete the exterior air barrier, seal all joints in the exterior framing with flexible caulk, construction adhesive, *elastomeric sprays,* or spray foam sealant. Seal key areas such as at joints in the sheathing, where top and bottom plates meet wall and floor sheathing, and between members in a double top plate. Some liquid-applied polymer products used for *flash sealing* require certified applicators. No matter which product you use, be sure to seal the joints, gaps, and cracks that are notorious for air leakage:

- Where two framing members abut (e.g., between the plates in a double top plate or between two adjacent studs in wall framing)

House wrap's primary purpose is water management. When properly installed, it also prevents air leakage from the exterior into the wall cavity. Although the Air Barrier Association of America mainly serves the commercial construction industry, the organization has a video online demonstrating proper installation of house wrap as an air barrier (http://tinyurl.com/housewrap).

- Where exterior sheathing and framing members intersect

- At joints in the sheathing

- Where the foundation and sill plate intersect (a sill gasket material between two elements will further protect this area)

- Where the bottom plate of exterior walls meets the subfloor

- Where the rim board meets the floor joist framing between all floor joists

- Where the bottom plate of walls meets the subfloor

- Between double top plates

- Between the rough opening and the window frame

- Between the garage and the living area (for IAQ and energy efficiency)

- Where any mechanical systems penetrate a wall, floor, or ceiling separating conditioned from unconditioned space

Interior Air Barriers

A home's construction should prevent indoor air from entering the wall cavity and the attic. Creating a barrier to prevent interior air from escaping is beneficial— especially in cold climates—because air entering wall cavities can filter through fibrous insulation, reducing its effective R-value and contributing to leakage to the outdoors. Moreover, in heating-dominated climates, if the relatively warm moist indoor air gets into wall cavities, it can condense inside the cooler wall cavity and create moisture problems. Interior air barriers aren't just good for home performance; they are essential for the tight construction the 2012 IRC requires, for the ENERGY STAR label, and to meet other programs' requirements.

Interior air barriers are as important for preventing airflow through the more obvious leakage sites (e.g., attic access panels) as for the less conspicuous ones (e.g., at drywall connections).

In extremely cold climates—those where homes don't have air-conditioning—the polyethylene vapor barrier can serve as an interior air barrier. In other climates, consult codes for the applicability of using vapor barriers or variable-permeability barriers, such as MemBrain.

CREATING A RAIN SCREEN

In marine climates or other areas where wind-driven rain is common, or where the house cladding is a water-absorbent material like stucco or stone, you can incorporate a *rain screen* in the siding attachment to protect the siding and the home from excessive moisture.

The idea is to create an air space between the exterior air/weather resistive barrier (e.g., house wrap or taped XPS foam sheathing) and the siding to reduce the pressure difference, which can drive rainwater behind the cladding. You can use a commercial product (e.g., Home Slicker) or attach furring strips vertically to the air- and weather-resistive barrier.

The air space not only reduces the air pressure difference across the siding, it creates an unobstructed path for rainwater to drain freely down the wall behind the siding. This can prevent entrapment of moisture and help wooden structures last longer. Finally, when sufficiently vented, the air space reduces the vapor pressure in the space, caused by sunshine falling on wet absorbent claddings, which can drive water vapor into a wall assembly.

ADDRESSING WHOLE-HOUSE ENERGY LOSS

Sealing rim joist and band joist areas can return more benefits than sealing other areas. Because of the typically low pressure at the bottom of a house (as warm air rises), the rim joist area above basements and crawl spaces contributes disproportionately to whole-house leakage, especially in cold climates. Cooler outdoor air is driven inward toward the low-pressure basement or crawl space through the rim joist area. Likewise, band joist areas between floors often contain ducts, so leakage causes disproportionate energy loss from ducts. When band joists are leaky, floors get cold and heating air becomes cooler.

To create an air barrier and insulate the wall cavity, some builders opt to fill the wall cavity with spray foam insulation. This method can provide an effective air barrier when properly detailed, but it is expensive. To complete an interior air barrier, wall cavity spray foam needs to be detailed at the ceiling plane (e.g., where drywall meets the top plate and at penetrations in the ceiling). To save money on insulation and still create a tight house, some builders do a flash seal with a layer of conventional spray foam insulation in the wall cavity or with a sealing-only product (such as Knauf's EcoSeal or Owens Corning's EnergyComplete) at key areas like framing junctures in the wall cavity. Afterward, they fill the wall cavity with conventional fiberglass or cellulose insulation. Elastomeric sprays can also help form the interior air barrier. ENERGY STAR considers open-cell foam at a thickness of 5.5" or greater and closed-cell foam at a thickness of 1.5" or greater to be an air barrier.

ENERGY STAR version 3 outlines climate-specific recommendations for air barriers. Review the ENERGY STAR Thermal Enclosure System Rater Checklist for air barrier recommendations.

In most homes and climates with an exterior air barrier, drywall forms the basis of the interior air barrier. Although drywall is a good air barrier, it requires proper detailing to prevent airflow through joints and at drywall connections. To create an interior air barrier using

USE VAPOR RETARDERS CAUTIOUSLY

You must be careful when constructing wall systems that employ vapor retarders.[25] As you improve the performance of wall systems by adding air barriers on the exterior, you create lower vapor permeability on the exterior (even when vapor-permeable materials are used) and the importance of vapor diffusion to the interior becomes more important for drying the assembly. Painted drywall is considered a Class III vapor retarder so vapor diffusion to the interior is already somewhat limited. Because the wall system is designed to prevent air exchange between the interior and the wall cavity (this air exchange is one method for drying old, leaky wall assemblies), more vapor diffusion to the interior must occur to dry assemblies that use vapor retarders. Section R702.7 of the 2012 IRC outlines conditions for using latex or enamel paint (*Class III vapor retarder*) instead of *Class I* or *Class II vapor retarders* in climate zones 5 and higher as well as marine 4. For best results, consult a design professional.

drywall, use gaskets, adhesive, or flexible spray-applied polymers (aka elastomeric sprays) to create a seal between framing and drywall at top and bottom plates, at window and door framing, as well as around electrical outlets, light switches, recessed lights, and at other areas that are typically leaky. Maintain continuity of the air barrier in areas that will not be covered in drywall (see Areas without Drywall in this chapter for more information).

Besides rim and band joist detailing and mechanical penetrations, a key area for preventing air exchange from the interior is at the top plate adjacent to attic spaces. This is particularly relevant in colder climates where the stack effect creates higher pressure that induces airflow towards the exterior at the ceiling plane. While exterior air barriers limit air leakage through the wall assembly, leakage to the attic can be addressed at the top plate. To seal this area, builders can choose from several methods that seal the drywall to the top plate. Some are sealing with elastomeric sprays or foam from the attic after the drywall is installed. Others are installing gaskets, adhesives, or elastomeric sprays (e.g., the EnergyComplete system) before or during drywall installation. Use air barrier techniques to prevent air infiltration through floors above garages (essential for IAQ and energy efficiency), above vented crawl spaces and unconditioned basements, and at cantilevered floors.

The Residential Diagnostics Database (http://resdb.lbl.gov/) includes *blower door* and other performance test results on more than 100,000 homes. A 2006 analysis of this data showed that homes that participate in an energy program such as ENERGY STAR are constructed 40% tighter than conventional homes.

Air barriers must be in full contact with the insulation. If they are not, temperature differences in a wall cavity can cause air currents to develop, reducing the wall system's insulating value. Aligning the air barrier and insulation isn't always simple.

Areas without Drywall

For continuity of the interior air barrier, install air barriers where there will not be a finished drywall covering. To prevent interior air infiltration into the wall cavity, install a rigid air barrier such as thin (composite material) structural sheathing, scrap drywall, wood sheathing material, foam sheathing, or spray foam insulation at all exterior walls behind fireplaces, tubs and showers, and stairs. The ENERGY STAR thermal bypass checklist can help you identify areas that require continuity. Other areas without drywall that require special detailing to maintain continuity of an air barrier include drop ceilings, knee walls, and cantilevered floors.

- **Drop Ceilings.** In drop ceilings and soffits (a solution for keeping ducts in conditioned space), a rigid air barrier at the ceiling plane will inhibit air exchange (i.e., create a pressure boundary) between the chase and the attic above. This rigid air barrier keeps any air that may leak from the ductwork in the house, rather than allowing it to leak into a vented attic where its energy is wasted.

- **Knee Walls.** At knee walls, air infiltration not only compromises the insulation (by wind washing), but air also can enter the floor cavity, making living space uncomfortable. Prevent this potential leakage by installing a rigid air barrier on the attic side of the knee wall, extending the air barrier to the subfloor, and sealing the top and bottom plates of the knee wall to preclude air infiltration around them (fig. 3.20).

- **Cantilevered Floors.**
  Cantilevered floors need an air barrier, not only to stop air from washing through the insulation in the cantilevered area, but also to prevent air from infiltrating the exterior wall and entering the floor joist area. For cantilevered floors, blocking (e.g., rigid foam or OSB) can be cut to fit between the floor joists and installed in line with the exterior wall plane. After air sealing around the blocking with spray foam or caulking, you can insulate the rim joist area with any type of

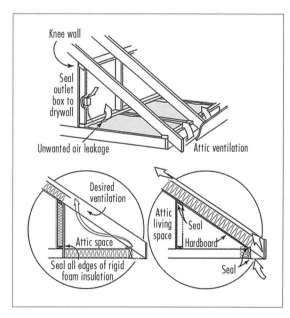

FIGURE 3.20 | **Knee wall air sealing**

Knee walls are a common source of air infiltration. *(U.S. DOE Energy Savers)* insulation. Other rigid materials should be used in the horizontal plane under the cantilevered joists to prevent wind-washing of the insulation in this overhanging area. Split-level homes built with a vented soffit covering fiberglass insulation (with the cantilevered floor joists creating an opening for air to infiltrate under the floor area) are drafty and waste energy (fig. 3.21).

FIGURE 3.21 | Cantilevered floor joist

A cantilevered floor joist area was covered with ventilated soffit, allowing airflow into the floor joist cavity and causing discomfort and excess energy use. Dark spots on the insulation reveal where air filtered through the insulation. *(NAHB Research Center)*

FIGURE 3.22 | Ventilation baffle

Baffles, whether commercially manufactured or hand built, transfer attic ventilation air to where it is needed—above the insulation. Sealing around these baffles prevents outdoor air from filtering through fibrous insulation. *(Jeannie Leggett Sikora)*

Vented Attics

In vented attics with soffit vents, baffles channel attic ventilation air to where it is needed (i.e., in the attic) and stop air infiltration through the attic insulation. For raised-heel and cantilevered roof trusses, rigid blocking over the top plate between each truss contains insulation and directs ventilation air above the insulation. Sealing the blocking material around its perimeter (and to adjacent ventilation baffles) can prevent air leakage through fibrous insulation (fig. 3.22).

INSULATION

The air barrier around a home prevents convective airflow losses whereas insulation is the blanket that reduces conduction losses. Insulation's R-value is a measure of its ability to resist heat flow across the wall cavity, from the warmer side of the cavity to the cooler one.

Fibrous Insulation

Fibrous insulation products like cellulose and fiberglass have a long history of blanketing American homes. Both are valid options for insulating green homes, but because both are air permeable, they must be coupled with air barriers.

Cellulose

Cellulose insulation contains up to 85% postconsumer recycled newsprint treated with non-toxic borates for fire and insect resistance. It can be used for wall or loose-fill attic insulation in new and existing homes. According to the Cellulose Insulation Manufacturers Association, it has the lowest embodied energy of any fibrous insulation. A skilled installer can fill a wall cavity completely with it, without gaps, voids, or compression. To earn points for green building certification and to get the highest performance out of the insulation, make sure you are getting a Grade I installation.[26] In a Grade I installation, insulation fills the wall cavity completely and uniformly and is enclosed on all six sides of the wall cavity.[27]

Because it is sprayed into wall cavities damp, the product needs sufficient time to dry before it is encased by drywall. Temperature and humidity impact drying time. You can run heaters, dehumidifiers, or both, to speed the process. Also, invest in an inexpensive moisture meter with a probe to measure moisture content in the insulation. The reading should be 19% or lower to prevent mold growth and wood decay.[28]

The interior air barrier will keep insulation dust out of the indoor air. Although cellulose insulation contains no additives that emit VOCs, some people are concerned about the inks in recycled newsprint emitting VOCs. To allay concerns, prevent a connection between the indoor air and the wall cavity with an interior air barrier.

Fiberglass

Fiberglass insulation options include

- kraft-paper-faced or unfaced batts;
- high density (provides a higher R-value within the same thickness as conventional fiberglass batts);

- blown in batts; and
- loose fill.

Blown-in-batt fiberglass insulation requires netting mounted to the face of studs to keep it in place. Many fiberglass products now have recycled content (usually 20%–30% postconsumer content). You can also get certified formaldehyde-free fiberglass insulation.

Other Fibrous Insulation

There are other insulation choices for green homes—mineral wool and recycled cotton. Neither is readily available for residential applications in most U.S. markets and the latter is more expensive than conventional products.

Mineral wool insulation contains postindustrial recycled material, is naturally fire and insect resistant, and maintains its R-value when wet. Recycled cotton insulation incorporates old jeans and contains 90% postconsumer waste.

As with any insulation product, refer to the codes for climate-based requirements.

Spray Foam

Spray foam insulation is commonly used because of its relatively high R-value and ability to seal and insulate in a single step. It is a popular way to seal band joists and create unvented attics. Products include open-cell and closed-cell spray polyurethane foams. Most are petrochemical-based products, but soy-based products are also available. Open-cell foam has a lower R-value per inch and higher vapor permeance. Both types may serve as air barriers when they are installed at the appropriate thickness (5" or more for open-cell foam, 1.5" or more for closed-cell foam).

MDI, which is one part of the two-part foam, is hazardous to the skin and when inhaled. Spray foam installers must wear protective equipment and respirators but the MDI dissipates quickly, especially in humid air, because it reacts with water. According to the *Spray Polyurethane Foam Association (SPFA)*, studies have shown MDI dissipates to safe levels in two to four hours. The SPFA suggests spraying at the end of the day and staying out of the home overnight. Both types of foam insulation are free of formaldehyde, and several manufacturers have received IAQ certification. Some open- and closed-cell spray foams include recycled content or incorporate renewable materials such as soy and castor oil in place of petrochemical-based oils. These spray foams still use MDI.

Special Considerations

Spray foam insulation can seal cracks and crevices because it expands after application. The chemical reaction responsible for the expansion also generates heat. Therefore, when they apply closed-cell or medium density spray foam over wires, applicators must install it in stages, spraying 1.5"–2" thickness and allowing it to cure before increasing thickness. At 2" thickness per pass, closed-cell foam temperatures can reach 150°F–180°F. At 4" or more, the foam temperature would be high enough to melt the plastic coating on wiring. Low-density (open-cell) foam temperatures during curing are lower so thicker passes are possible. Finally, if installers cover junction boxes with foam, the boxes should be clearly marked to facilitate any future electrical work.

Rigid Foam

In conventional light-framed construction, insulation is placed only in wall cavities. Since wood has an R-value of about 1 per inch, a 2 × 6 stud will have an R-value of less than 6, whereas the wall cavity insulation will be about R-19. If studs are spaced 16" o.c., about 15% of the wall surface area will have an R-value of 5.5 instead of R-19, which reduces

IRC Wall Bracing: A Guide for Builders, Designers, and Plan Reviewers (http://www.foamsheathing.org/IRC_Wall_Bracing_Guide.pdf) and the *Guide to Insulating Sheathing*, available from http://www.buildingscience.com, discuss how to use insulating foam as exterior sheathing. The Foam Sheathing Coalition, an industry organization that was established in part to promote the proper use of foam sheathing in the construction industry, contains technical support information for all types of foam sheathing on its website, http://www.foamsheathing.org (click "technical" on the left navigation bar).

the overall effective R-value of the wall system.

Continuous insulation, such as exterior rigid foam sheathing, can improve the wall's thermal performance and reduce the *thermal bridges* created by the studs.

Rigid foam insulation—which can be *extruded polystyrene (XPS)*, *expanded polystyrene (EPS)*, or *polyisocyanurate (polyiso)*—applied to the exterior of a wood-framed home is beneficial for several reasons. It provides a continuous layer of insulation and an exterior air and water barrier when seams are taped and the foam is integrated with the rest of the home's flashing system.

Each type of rigid foam sheathing has its own strengths and application in green home construction.

- **XPS foam.** Some XPS foam products contain postindustrial recycled content, but none currently has postconsumer content. XPS foam is recyclable, although finding local recycling outlets can be a challenge. With its high compressive strength and resistance to moisture absorption, XPS foam is suitable for below-grade applications where the code permits. The Extruded Polystyrene Association (http://www.xpsa.com) has technical resources for using XPS. Click "technical information" on the navigation bar.

- **EPS foam.** EPS is common in ICFs and SIPs. It is recyclable but, as with XPS foam, it is difficult to find recyclers. It is the most fragile of the rigid foams, the least expensive, and the most vapor permeable. Because it is susceptible to damage, it is not often used as a sheathing product. It is HCFC free (only non-ozone depleting chemicals are used in manufacturing it). It can absorb more water than XPS foam insulation and is not typically used below grade (except in water-managed ICF foundations). The EPS Molders Association, which promotes using EPS foam sheathing in construction, offers technical and green building information on its website (http://www.epsmolders.org). Click "green building" on the upper navigation bar.

- **Polyiso foam.** Foil-faced, polyisocyanurate boards are HCFC-free. Virtually all manufacturers use recycled content—which varies by manufacturer and product but can be up to 100%. Its moisture-absorbing properties make it unsuitable for below grade applications. Furthermore, protection from moisture during storage and handling is important.

Structural Foam Sheathing

A hybrid type of sheathing product, structural foam sheathing combines the structural properties of wood panels with the insulating ability of rigid foam sheathing. A few types are on the market and the Foam Sheathing Coalition anticipates more manufacturers will get on board in the future. Each product has unique properties and applications.

The American Chemistry Council (http://www.americanchemistry.com) has information on where to recycle plastics such as foam sheathing, scrap plumbing pipe, and siding. Click on "environment" under "innovation" in the upper navigation bar and select "recycled plastics markets database."

PROVEN PRACTICES

Throughout the green home construction process, from digging the foundation to nailing the last roofing shingle, there are proven best practices for lessening the project's environmental impact. As you incorporate green construction practices, think about how you can save natural resources and reduce waste while creating a house that is built to last. Many of these green practices eliminate waste, improve comfort and durability, and can save you and the home owner money.

Guide to Insulating Sheathing (http://www.buildingscience.com) provides a thorough overview of insulating foam sheathing in residential construction.

4

HEATING, VENTILATION, AND AIR CONDITIONING

Careful design and selection of energy-efficient heating, ventilation, and air-conditioning (HVAC) systems helps control interior temperature, relative humidity, and IAQ, contributing greatly to a more comfortable indoor environment and to building durability. Although high performance homes may generate smaller heating and cooling loads, they also are designed to maintain more precise indoor environmental conditions than other homes. Therefore, the HVAC systems in high performance homes must be carefully planned, installed, and operated.

MECHANICAL VENTILATION

Mechanical ventilation is the process of removing pollutants at their source through spot exhaust ventilation and introducing a controlled amount of fresh air into a house to dilute indoor air pollutants and promote good IAQ. Various options are available for mechanical ventilation. Local climate, builder preference, cost, practicality, and house construction influence system selection.

Residential ventilation, beyond kitchen and bath exhaust ventilation, is still evolving in the United States. Although some states have mandated residential ventilation for years, builder-specific learning materials for the design and installation of residential ventilation systems are limited. Moreover, the few residential ventilation experts that exist don't always agree on the best way to provide ventilation. One thing they agree on, however, is that mechanical ventilation is essential for tightly constructed green homes.

Spot (Local Exhaust) Ventilation

Removing pollutants and moisture at the source is the first step in ensuring good IAQ in high performance homes. Spot exhaust ventilation is essential for removing pollutants, moisture, and odors before they have a chance to disperse throughout the home. Home owners must operate exhaust ventilation systems and you can encourage them to do so by installing controls such as timers or humidistats. You can also inform home owners about the proper operation and the importance of local exhaust ventilation. Follow guidelines regarding local exhaust fan capacity from organizations such as the Home Ventilating Institute (HVI).

Kitchens

Ventilate kitchens in green homes as follows:

- Always duct the exhaust to the outdoors and flash the opening.
- Don't oversize the fan, especially in a tightly built house.
- Select ENERGY STAR range hood fans that are rated for efficiency and noise. (Consider multispeed fan control to keep noise low when you don't require a high volume of airflow.)
- Instruct home owners on the proper operation of spot ventilation to remove pollutants at their source.

HVI offers recommended ventilation rates for range hoods, which vary based on the location of the range (table 4.1).

TABLE 4.1 | HVI-recommended ventilation rates for range hoods

RANGE SIZE AND POSITION	30"	36"
Along wall	250 cfm (100 cfm minimum)	300 cfm (120 minimum)
Above island	375 cfm (125 minimum)	450 cfm (150 minimum)

Bathrooms

To select an energy-efficient bathroom fan, determine the fan's capacity based on HVI (http://www.hvi.org) recommendations (table 4.2).[29]

TABLE 4.2 | HVI-recommended bathroom ventilation rates

BATHROOM SIZE	MINIMUM VENTILATION CAPACITY
<50 sq. ft.	50 cfm
50 to 100 sq. ft.	1 cfm per sq. ft. bathroom area
>100 sq. ft.	50 cfm per fixture (toilet, shower, bathtub, jetted tub)

Then you can use the ENERGY STAR ventilation fan products list (http:// tinyurl.com/ChooseAFan)

to find one that produces a high CFM per watt at a low sone level. You can narrow the list by manufacturer.

Whole-House Mechanical Ventilation Systems

Tightly constructed homes, by design, have little natural air leakage so they rely on mechanical ventilation to bring fresh air indoors. While this fresh air often requires energy to condition it (the "energy penalty" associated with ventilation), it is less expensive than heating and cooling the uncontrolled air that leaks into and out of a home that is not tightly built (and fixing the moisture issues the leakage can cause). Some ventilation systems can recover energy to preheat or precool the incoming airstream.

The amount of fresh air to bring into a home is a key HVAC design question: too much air requires excessive energy usage for ventilation fans and for space conditioning, but too little may increase humidity or make a home feel stuffy. Although building scientists debate how much ventilation a structure needs to provide high-quality indoor air without dramatically sacrificing energy efficiency, the American Society of Heating and Refrigeration Engineers Standard 62.2 (ASHRAE 62.2) has general guidelines for residential ventilation. Some states and localities set their own requirements.

You can determine the recommended continuous residential ventilation rate, based on ASHRAE 62.2-2010, using the following equation:

$$\text{continuous ventilation rate} = 7.5 \text{ cfm} \times (\text{number of bedrooms} + 1) + 0.01 \text{ cfm} \times (\text{s.f. of conditioned space})$$

Ventilation air can be (and often is) drawn in intermittently, but when averaged, the intermittent ventilation rate should equal the continuous rate. In other words, if you are planning to operate the ventilation system 20 minutes per hour, select a

VENTILATION VS. LEAKAGE

In the past, leaky homes exchanged a lot of air with the outdoors through natural infiltration. But even in leaky homes, natural infiltration rates vary with outdoor temperature (the most leakage often coinciding with the biggest heating loads) and with other factors that create a pressure difference between indoor and outdoor air. Building a home tightly and installing a mechanical ventilation system is the best way to get consistent and recommended ventilation rates.

ventilation system with the capacity to introduce air at three times the recommended continuous rate during those 20 minutes (60 minutes ÷ 20 minutes = 3).

Mechanical ventilation encompasses a variety of systems that bring fresh outdoor air into a home to replace stale air. There are many options for designing a system to draw, distribute, and control the amount of ventilation.[30]

First, decide how fresh air will be distributed around the home, including whether to include separate ventilation ducts or draw fresh air through the heating and cooling ductwork. Most systems use the home's existing ductwork, but the paradigm may be shifting toward dedicated ventilation air distribution systems in high performance homes.

In the ideal residential mechanical ventilation system, ventilation air enters and exits through—depending on climate—either a heat-recovery or an energy-recovery system. In this ideal system, ventilation air travels through separate ducts throughout the home, ensuring airflow goes precisely where it is needed. Separately ducting ventilation air, however, is not always the most practical approach to residential ventilation.

Here are some guidelines for distributing fresh air using a separately ducted system, with or without heat recovery:

- Provide a supply duct in each room and common area and locate registers close to the ceiling with airflow directed at the ceiling.

- Install return outlets in each high-moisture area such as the kitchen (on an interior wall at least 10' from the oven or cooktop), utility room, and bathroom.

- Follow the manufacturer's limits on duct length and keep runs as smooth and straight as possible.

In the real world, it is more common to see a system that uses the home's central air handler and ducts to circulate ventilation air. It costs less, is easy to design and install, and it filters, conditions, and distributes ventilation air. The system often has a direct air inlet vent that feeds into the return side of the air handler. A motorized damper with an electronic control prevents overventilation. As ventilation air is introduced through the inlet vent and circulated through the home by the central air handler, the house becomes slightly pressurized and exhaust air flows out through the path of least resistance (such as through exhaust fan ports). Some builders install dedicated central exhaust fans to balance flow.

The system works well enough (particularly for warm, humid climates where house pressurization is more desirable than depressurization), but it is expensive to operate because the air handler is oversized for the job of ventilation. It draws 250 watts, give or take, with an efficient motor, and twice that with a conventional motor.[31] Also, the system can't capture energy from the exhaust air, which flows freely out of the house. By pressurizing the house, the supply-only system inhibits pollutants such as radon or garage fumes from being drawn inside (fig. 4.1). However, in cold climates pressurization can push relatively warm moist indoor air out through the building walls and ceiling, where it can condense on the cold surfaces. Building Science Corporation[32] suggests incorporating a continuously operating exhaust fan in a central-fan-integrated system for a very cold climate. Interlocking the controls for the exhaust, supply, and air handler fans in this type of system encourages balanced airflow and optimum fresh air distribution.[33]

Another way to use the central air handler for balanced ventilation in high performance homes in any climate is to add an HRV (fig. 4.2) or an ERV. An HRV simultaneously delivers preheated (or precooled) air to the return duct system while it draws equivalent air volume to the outdoors. This type of system synchronizes the air handler control

VENTILATION SYSTEM ENERGY RATINGS

When this book was published, there were no ENERGY STAR-labeled *heat recovery ventilators (HRVs)* or *energy recovery ventilators (ERVs)*. However, Canada has ENERGY STAR standards for HRVs and ERVs and maintains a database of ENERGY STAR products in Natural Resources Canada's EnerGuide (http://oee.nrcan. gc.ca/energuide). Another option is to use the Home Ventilating Institute's (http://www.hvi.org) directory of certified products. It shows sone level, airflow, fan power usage, and energy recovery efficiency for HVI-certified HRV and ERV equipment. Besides the Canadian EnerGuide, the HVI directory may be your best bet for comparing HRV and ERV equipment efficiency.

with the HRV or ERV using a timer or more sophisticated control. Whenever the HRV or ERV is operating, the air handler also operates. The ventilation load (which is a function of climate and other factors like equipment efficiency) determines the estimated energy savings from using an HRV or ERV. Although in some cases HRVs and ERVs may not be economically justified by energy savings, they may perform better and make a space more comfortable. They are more efficient because they capture most of the energy in the exhaust airstream. For optimum performance, the intake and exhaust airstreams in these systems must be balanced using flow gauges and balancing dampers.

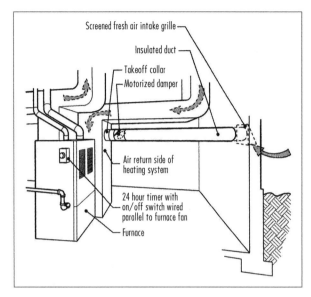

FIGURE 4.1 | **Typical supply-only system**

A typical supply-only central-fan-integrated ventilation system pressurizes a house but you can add relief vents to reduce the pressure difference between the home and the outdoors. *(Washington State University)*

Builders in some regions install continuously operating, low-wattage exhaust fans, such as those typically installed in bathrooms, with passive makeup air (an exhaust-only system) or with a balanced, fan-powered

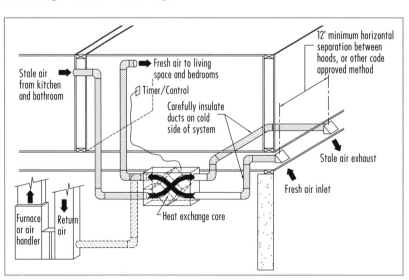

FIGURE 4.2 | **Basic HRV**

As it replaces stale air from a house with fresh air from outdoors, the heat exchanger core in an HRV preheats or precools the incoming airstream. *(NAHB Research Center)*

makeup air supply (a balanced system, as figure 4.3 shows). In exhaust-only systems, infiltration, individual fresh air ports, or trickle vents supply makeup air.

Although ventilation exhaust air may be drawn from the kitchen area, it should never be drawn near a range hood. Keep ventilation returns at least 10' from a range or oven.

Exhaust-only systems have potential problems including the following:

- They may make the inside air uncomfortable, particularly near inlet ports and in cold climates.

- They do not recover lost heat.

Exhaust-only systems with dedicated ports or inlets for makeup air can make the inside air uncomfortable, particularly near inlets and in cold climates.

An exhaust-only system that relies on natural infiltration for makeup air supply has other potential issues:

- The house may become depressurized. This induces unplanned air infiltration through gaps and cracks in the construction if there is more exhaust airflow than intake airflow.

- As the house becomes depressurized, underground radon could seep into a basement or crawl space.

- In warm humid conditions, outdoor air drawn into the building envelope can condense on relatively cool interior walls, causing moisture problems within the wall cavity.

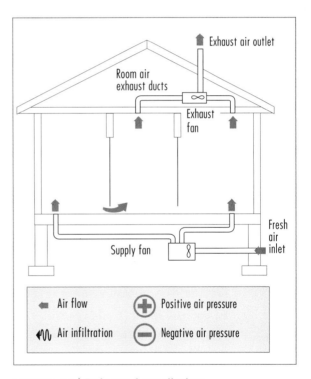

FIGURE 4.3 | **Balanced ventilation**

This type of balanced ventilation system relies on exhaust fans and powered supply fans to avoid creating a pressure difference between the home and the outdoors. The system design must address where to deliver the unconditioned ventilation air so it does not make a space uncomfortable. *(U.S. DOE)*

LOCATE HRVS AND ERVS IN CONDITIONED SPACE

For best performance, HRVs and ERVs should be located in conditioned space, fitted with a condensate drain, and include intakes to draw fresh air. An intake should not be located near an exhaust vent or anywhere the air could be polluted. The installation contractor should measure and balance intake and exhaust airflows. As with any building product, always follow the code and manufacturer's installation guidelines precisely so that systems meet performance expectations.

Where to introduce the makeup air supply is a chief design consideration for exhaust-only systems with dedicated makeup air to keep occupants comfortable and ventilation air well distributed.

There are many other methods for introducing ventilation air, distributing it around a home, and controlling a ventilation system. Each has its own design considerations, depending on climate and house construction. *Ventilation Guide*[34] and *Residential Ventilation Handbook*[35] review the options thoroughly. An online guide, *Review of Residential Ventilation Technologies* (http://www.buildingscience.com) is also available. Click "information" and search by title.

NON-MECHANICAL VENTILATION

There are natural and passive ventilation methods to bring fresh air indoors without mechanical energy and without opening windows, which is inefficient during heating and cooling seasons. Unfortunately, although these methods are popular in Europe, they require design features—such as towers that promote the *chimney effect*—more suited to custom than production homes. Moreover, although these systems can be a legitimate component of a green home, they do not provide reliable and consistent levels of ventilation, cannot filter incoming air, and are not universally applicable.

INSTALLING VENTILATION SYSTEMS

Careful ventilation system installation is key to getting the best possible performance out of the system. *The Ventilation Guide*[36] referenced previously recommends the following practices:

- Keep ductwork as straight as possible to prevent airflow constriction.

- Consider where condensation may occur (where a relatively warm and moist airstream meets a cold surface) and insulate ductwork accordingly.

- If you are using the central duct system to distribute ventilation air, be sure it doesn't exhaust fresh air before the air is distributed throughout the home (called short-cycling). You can introduce fresh air into the HVAC system return duct and have a separate central exhaust duct independent of the HVAC system to balance air pressure. Ideally the central exhaust fan operates concurrent with the supply air fan.

HVAC CONTRACTORS

Depending on your local market, finding a company that is knowledgeable about residential ventilation system design, installation, and control can be difficult. The Heating, Refrigeration, and Air Conditioning Institute of Canada (HRAI) offers residential ventilation design and installation certification— the HVAC training group RSES (http://www.rses.org) and other U.S. organizations offer training. Two U.S. organizations, North American Technician Excellence (NATE) and Building Performance Institute (BPI), offer continuing education credits so the HRAI's database of certified U.S. contractors (http://www.hrai.ca/certification_us.php) will continue to expand. In the future, look for online training and, possibly, new residential ventilation certification programs from U.S.-based organizations.

- Ensure the ventilation system distributes fresh air evenly throughout the home via dedicated ventilation ducts that deliver air to each room or by using the central fan for circulation.

- Unconditioned ventilation supply air can make an area feel drafty. Therefore, avoid placing supply outlets for a dedicated ventilation air distribution system where the air might blow directly on people. Also consider using multiple supply outlets in a single room to reduce draftiness near outlets.

- Draw ventilation air from areas where the air is cleanest (e.g., keep inlets away from combustion appliance exhaust vents, driveways, garages, rooftops, cooking grills, and outdoor fireplaces). Consult codes for appropriate inlet locations.

- After installation, verify (using equipment like a *flow hood* or an *anemometer*) that the fan flow meets design specifications.
- Select equipment with a low sone rating so it will not create unacceptable noise and prompt occupants to turn it off.

ALTERNATING SUPPLY AND EXHAUST VENTILATION

Yet another option for ventilating homes is to provide a semibalanced ventilation system that relies on intermittent supply and intermittent exhaust to produce balanced ventilation over time. These combination systems will not cause the home to be pressurized or depressurized over long periods of time. One option for avoiding long-term pressurization in a cold climate is to couple a CFI supply-only ventilation system with continuous exhaust ventilation.

Air pollutants may enter homes through attached garages. An air barrier between the garage and attached living area is essential for stemming the flow of pollutants from the garage. Some builders add spot exhaust ventilation systems to attached garages. These operate continuously on timers or for several minutes after the garage door closes. Other builders simply construct detached garages.

WHOLE-HOUSE FANS VS. VENTILATION SYSTEMS

Whole-house ventilation systems are designed to improve IAQ, not for temperature control (unlike whole-house fans designed to rapidly draw in outdoor air for cooling). However, some whole-house ventilation systems can operate on an "economizer" cycle and draw air in for cooling. These systems sense outdoor temperature and relative humidity and, when conditions are suitable, draw in relatively cool outdoor air (e.g., at night) to temporarily reduce or eliminate the need for mechanical cooling.

HEATING AND COOLING DISTRIBUTION

Along with higher efficiency equipment, tight duct design and meticulous installation techniques have improved heating and cooling system performance. Whereas 25% of conditioned air might leak from older air distribution systems, today's well-designed and installed ducts leak 6% or less supply air. But energy loss from leaky ducts isn't the only performance issue: leaky return ducts can draw air into the duct system (often from the basement). This condition increases pressure differences between indoors and outside and reduces home performance. It can induce greater whole-house leakage or, worse, suck radon into the duct system and distribute it around the house.

When all ductwork is in conditioned space and the house is well sealed, duct leakage provides useful energy to the house, rather than wasting this energy to the outdoors. Yet even when ducts are in conditioned space, tight duct systems minimize pressure imbalances and distribute air as they were designed to, keeping room temperatures consistent and occupants comfortable. In general, using industry-accepted design procedures, following general installation guidelines, and testing and balancing airflow after installation will optimize duct performance.

Backdrafting is a potentially dangerous condition through which combustion by-products enter a home. In tightly constructed homes where makeup air for combustion is scarce and backdrafting may occur, builders should always specify sealed combustion appliances (e.g., fireplaces, water heaters, and furnaces) and hard-wired CO detectors.

Duct Design and Installation

To meet the green building project goals of energy efficiency, comfort, and durability, apply industry and manufacturer standards for designing and installing HVAC systems. A well-designed and carefully constructed HVAC system featuring high-efficiency equipment, which is installed and tested according to industry protocols, will meet the goals.

LOAD CALCULATIONS AND DESIGN

The first step to creating a high performance air distribution system for a home is to hire an HVAC contractor that uses software to calculate heating and cooling loads according to the most recent version of *ACCA Manual J Residential Load Calculation*[37] (version 8 or later). These predictions will only be as good as the information entered. Therefore, the contractor will need precise and complete information about the home's construction. Provide the contractor with the following specifications:

- Wall system
- Roof system
- Window overhangs
- House orientation
- Window and door properties (e.g., U-factor, SHGC)

You can't predict whole-house and duct air leakage before a home is completed, but you can make an educated guess based on previously constructed homes that used the same specifications.

The ACCA Manual J calculations for each room are the cornerstone of a high performance HVAC design. The calculations consider insulation and air-sealing measures, anticipated internal loads, and solar heat gain. A diligent contractor will strive to get accurate details about the construction prior to making calculations. Don't use contractors who rely on rules of thumb or calculate loads on the back of an envelope.

EQUIPMENT SELECTION

ACCA Manual S addresses equipment sizing. Building scientists have long touted the benefits of right-sizing equipment (especially cooling equipment) over the common practice of oversizing. Multi-stage compressors and variable speed air handlers can rectify some of the performance problems oversizing causes (e.g., lack of humidity control, excessive energy use).[38] Nevertheless, solving those problems does not overcome the higher price tag for oversized equipment. Hence, right-sizing is a smart choice but it requires careful design, operation, and equipment selection.

AIR DISTRIBUTION AND DUCT DESIGN

After you select the HVAC equipment, you can use its specifications and two ACCA manuals to design the rest of the air delivery system. *ACCA Manual D—Duct Design*[39] guides the design of the ducts to deliver the right amount of air to and from each room (to meet the loads calculated by Manual J). The duct design depends on the performance of the air handler you select. Follow sound design principles (i.e., keeping ducts as short and straight as possible) to get the proper airflow to each room. *ACCA Manual T—Air Distribution Basics*[40] is a resource for selecting and placing grilles and registers to distribute the air around a room.

Return and Supply Airflow

A home should have balanced return and supply airflow for several reasons. On a whole-house scale, unbalanced flows can pressurize or depressurize a home, exacerbating whole-house air leakage. Equivalent pressure also affects room-to-room comfort: higher pressure on one side of a doorway than on the other can constrict supply airflow, creating uneven temperatures and stagnant air.

For return airflow, many high performance builders are combining *transfer grilles* (fig. 4.4) or *jump ducts* across doorways with centrally located returns. These systems cost less and use fewer materials than having dedicated returns to each room (or relying on inadequately sized door undercuts that impede airflow). Nonetheless, they distribute air effectively around the home.

FIGURE 4.4 | Transfer grille

A rough in for a transfer grille that will provide a return air pathway and balance air pressure across the doorway. *(Jeannie Leggett Sikora)*

Designing air ducts using the four ACCA manuals generates a plan for delivering the right amount of air throughout a home but overall system performance depends on duct installation according to the design. An HVAC system can perform to its maximum potential for comfort and energy usage only if it is installed precisely according to design and ducts are carefully sealed against air leakage.

QUALITY INSTALLATION

Follow these general principles for installing ductwork:

- Keep runs as short and straight as possible.
- Avoid unnecessary bends and elbows which impede airflow.
- Don't run ducts through unconditioned space (e.g., use drop ceilings in single-story homes or bury well-sealed ducts under insulation).
- Don't compress or constrict airflow in flexible ducts.
- Fully extend all *flex duct* so it doesn't sag.
- Provide dampers at take-offs, where branches meet the main trunk line, to effectively balance airflow.
- Never use panned joist cavities; they are notoriously leaky.

ACCA Standard 5 (*HVAC Quality Installation Specification*)[41] is the industry standard for quality HVAC installations. This standard, (https://www.acca.org/Files/?id=116) is a valuable guide for a green builder in developing HVAC trade contractor scopes of work. However, some criteria in Standard 5 (e.g., total duct leakage) may differ from green building program criteria, so ensure your trades are meeting the most rigorous requirements. Keep in mind that version 3 of ENERGY STAR requires using contractors who are Quality Assured through the ACCA or a similar program.

HVAC Quality Installation Guidebook, a compendium to the ENERGY STAR HVAC System Quality Installation Contractor Checklist, contains photos and tips for essentials like avoiding compression in flex ducts, providing airflow dampers to critical sealing points, and balancing pressure across doorways. Go to http://www.energystar.gov and search the guidebook's title.

Sometimes a home's framing or other factors impede perfect installation. For example, a floor framing member may line up directly under the interior first-story wall that you intended to run ducts through to get to the second story, preventing a straight run through that wall. If you decide where to place the mechanicals during the design phase, and plan the framing accordingly, you can avoid problems like this.

High-Efficiency Air Filters

Forced air systems stir up and will circulate particulates throughout a home. To remove contaminants (and to help protect the air handler and coil), use high-efficiency filters and instruct home owners to do the same.

Most green building programs offer points for using filters with ratings of MERV 8 or higher. MERV 8 indicates the filter will capture 70% of the particles between 3 and 10 microns in size at a standard test airflow rate. At MERV 9, the filter will trap 85% of the particles that MERV 8 captures, plus half of the particles between 1 and 3 microns. As MERV increases, the percentage of particles of a certain size that the filter captures increases. However, the pressure drop across the filter also increases. To account for the pressure drop in the system and ensure the air handler fan can still push air around the entire house, an HVAC designer needs to know what type of filter you will install. Higher MERV filters get clogged more quickly than do less efficient filters. Therefore, your home owner's manual should instruct the owners about filter maintenance and replacement. *Your New Green Home and How to Take Care of It: Homeowner Education Manual Template*[42] provides an efficient way for you to create a green home manual for owners.

RADIANT DISTRIBUTION

Radiant heat is another option for heating new green homes. It generally adds to the price of a new green home but some buyers are willing to pay for its benefits. Radiant distribution, whether via in-floor or baseboard radiators, is quiet, helps maintain comfortable wintertime indoor relative humidity levels, and, when properly insulated, limits energy losses associated with distribution.

Radiant heat distribution generally garners credit from green building rating programs because it can enhance IAQ and energy efficiency. These systems don't stir up dust and particulates; creating multiple, separately controlled zones is simple; and energy losses associated with distribution can be minimal. Most home owners find the systems comfortable because they distribute heat evenly and don't create excessively dry air indoors in the heating season.

Radiant systems are rarely used for cooling because cool surfaces are often below the dew point of air, causing condensation. When you use radiant heating, you will likely need a separately ducted cooling system. Therefore, you will need to purchase more materials and incur other additional costs.

As with other types of systems, you must carefully design and install a radiant heating system to manage costs and optimize performance. The Radiant Panel Association (RPA) certifies designers and installers and radiant distribution product manufacturers offer fee-based design services.

Beyond selecting an energy-efficient boiler or furnace to heat the fluid, four other factors contribute to system efficiency: supply water temperature (reducing this typically increases efficiency), temperature drop through system (increasing this can reduce pumping energy), variable speed pumping (to match energy for pumping with the heating load), and preventive maintenance.

The Radiant Panel Association (RPA) maintains a directory of certified *hydronic* systems designers and installers (http://www.radiantpanelassociation.org). Click "Find a Contractor" and select "Certified Only" to find certified designers or certified installers (separate certifications).

HEATING AND COOLING EQUIPMENT

The right heating and cooling equipment complements thoughtful design and meticulous installation. Options for energy-efficient heating and cooling equipment are abundant.

Heating Equipment

Begin narrowing equipment options by considering available fuel. Natural gas is a logical choice where it is available because it is relatively inexpensive and efficient. You can also choose dual-fuel systems that use electricity or gas depending on weather conditions.

ENERGY STAR products are the minimum threshold for a green home. The ENERGY STAR label requires Annual Fuel Utilization Efficiency (AFUE) of 90%, meaning that 90% of the fuel is converted to heat energy. However, manufacturers offer gas furnaces with up to 98% AFUE.

For gas equipment, look for condensing, sealed combustion burners which can modulate the output to meet the load. These systems will prevent short-cycling and promote comfort, efficiency, durability, and IAQ.

Cooling Equipment

Conventional air-conditioning equipment includes an outdoor compressor, an indoor heat exchanger coil, and an air handler to distribute the air. Air conditioners have become more environmentally friendly since ozone-depleting refrigerants were phased out and Seasonal Energy Efficiency Ratings (SEER) were raised to 20 and above (the federal minimum is SEER 13).

The biggest advance in air-conditioning efficiency is the variable capacity system. Air conditioners that are sized for design loads (which represent relatively extreme conditions) can meet cooling loads rather quickly under most conditions, resulting in frequent cycling during less extreme weather conditions. With multi-stage compressors and variable-speed blowers, air conditioners can meet loads while operating at a lower capacity for longer periods of time. Continuous operation is the most efficient way for equipment to run. With today's high performance homes, selecting equipment that meets the latent and sensible loads is more critical than ever.

HEATING AND COOLING EQUIPMENT CERTIFICATION

The Air-Conditioning, Heating, and Refrigeration Institute (AHRI) certifies the performance of heating and cooling systems. Most green building programs require an AHRI certificate for air conditioners and heat pumps. The certificate ensures that the indoor and outdoor components are compatible; unmatched components reduce system efficiency. System efficiency ratings are listed at http://www.ahridirectory.org.

Evaporative Cooling

Evaporative cooling is an efficient air-conditioning option for hot, dry regions. Evaporative coolers, also known as "swamp" coolers, reduce air temperature by running a dry air stream through a water-soaked pad. As the water evaporates, it absorbs some of the energy in the air, reducing the air's temperature (*sensible heat*) and increasing humidity (*latent heat*). Indirect systems use the same principle, coupled with a heat exchanger, so moist air is not circulated directly through the house. Both systems rely on a dry airstream to reduce the temperature. Therefore, they are not suitable for mixed or humid climates.

Absorption Cooling

Some air conditioners rely on thermal energy to drive the absorption refrigeration cycle. Although using thermal energy is appealing— particularly when the heat source is solar energy—and despite research and development efforts, this option is not yet available for residential use in the United States.[43]

Air Source Heat Pumps

Air source heat pump technology has also advanced in recent years to incorporate thermostatic expansion valves (TXV) for precise refrigerant control, variable speed blowers, multi-stage compressors, and improved heat exchange coil design. Heat pumps are available with SEER ratings of more than 20 (federal minimum is currently SEER 13). With one piece of equipment providing both heating and cooling, air source heat pumps can be an economical choice for a builder, especially in an all-electric home.

If you are building in cold climates, reverse cycle chillers are an option. Reverse cycle chillers are a twist on the air source heat pump that heats or cools water or air for radiant heating, air conditioning, and domestic hot water production. The systems store energy in a tank of hot water to add flexibility and to resolve some of the performance issues with inadequate air delivery temperature in very cold weather (especially during the defrost cycle). Reverse cycle chillers rely on the stored hot water for defrost. They are compatible with hydronic distribution. In the cooling season, the storage tank can collect excess thermal energy for domestic water heating just like a geothermal heat pump *de-superheater* does.

Equipment Ratings

You can research the comparative energy efficiency of heating and cooling products using two sources, the Consortium for Energy

Efficiency (CEE) Directory of Energy Efficient HVAC Equipment (http://www.ceedirectory.org) and the ENERGY STAR website (http://www.energystar.gov).

Some CEE program criteria go beyond ENERGY STAR requirements. For example, gas heating equipment meeting CEE's Tier 3 (most efficient) specs must have an AFUE of 94 (ENERGY STAR requires AFUE 90) and electricity can comprise no more than 2% of total energy consumption. The criteria considers blower motor efficiency, unlike ENERGY STAR criteria.

To find the highest efficiency equipment rated by CEE, select the type of equipment you are seeking from the home page. After you get to the AHRI directory, select CEE Tier 1, 2, or 3 (the most efficient) from the drop-down menu or narrow your search by manufacturer, SEER rating, or other criteria. You can download AHRI certificates, which you usually must submit to green building rating verifiers, from the directory.

For a list of heating and cooling products that meet ENERGY STAR criteria, click "Find Products," select the type of product you are seeking, and then click "Qualified Products" to download a list. These lists don't always include capacity and other information but they are one tool to compare the efficiency of available products.

You may want to include a home automation professional in your green home projects. New heating and cooling technology allows users not only to program temperature, but also to control humidity and ventilation. The Custom Electronic Design and Installation Association (CEDIA, http://www.cedia.net) offers a number of professional certifications related to home automation and a directory of its certified professionals.

Ground Source Heat Pumps

Ground source heat pumps provide the highest efficiency of any electric heat pump option. In general, well-designed and properly installed *geoexchange* or Ground Source Heat Pumps (GSHPs) operate approximately 50% more efficiently than similarly designed and installed air source systems. In addition to being efficient, they are often desirable because there is no outdoor compressor unit (with its associated noise). Instead, they have an underground loop that moves heat from the ground to the house (or vice versa). Unfortunately, GSHPs are comparatively expensive so they have been limited largely to high-end or multifamily homes. However, specific designs can reduce costs. For example, you can install a community ground loop instead of multiple individual loops to improve affordability.

To ensure maximum performance of a GSHP, the pump and its underground loop must be designed and installed carefully. The International Ground Source Heat Pump Association (ISGHPA) certifies GSHP designers and installers. To find a certified geoexchange designer, certified installer, or accredited loop installer, go to http://www.igshpa.okstate.edu. Manufacturers also may certify designers and installers. Figure 4.5 shows the well-drilling process for a vertical loop installation.

FIGURE 4.5 | GSHP well drilling

Ground source heat pumps exchange energy with the ground through an underground loop. Here, a well driller begins the process of installing a vertical loop. *(Jeannie Leggett Sikora)*

When specifying GSHP in cooling-dominated climates, you can add a de-superheater that recovers hot water from the waste heat created by the heat pump and increases cooling efficiency.

Ductless Mini-Split Heat Pumps

Another category of heat pumps is the ductless mini-split. These systems pipe refrigerant to separate air handler coils located in each heating and cooling zone, rather than having a central cooling and heating coil with a duct system to distribute air. These through-the-wall air handlers resemble low-profile window air conditioners. Mini-split air heat pumps eliminate energy losses associated with ductwork, have EER ratings comparable to conventional packaged equipment, and use less interior space for distribution.

Dual Fuel Systems

Dual-fuel heat pumps can heat a home with gas or electricity, depending on outdoor conditions. In these systems, the heat pump serves as the primary source of heating and the gas furnace kicks in when ambient temperatures go below a certain point. This increases system efficiency and boosts the delivery temperature as necessary. These systems eliminate the need for electric resistance auxiliary heat and, therefore, cost less to operate than conventional heat pumps.

ENLISTING CUTTING-EDGE PROS FOR HVAC WORK

Planning, designing, and installing HVAC systems for green homes can often be the most complex aspect of going green. To ensure the HVAC system—and a new green home—perform as they were designed to, enlist experienced contractors who are willing to learn the building science behind HVAC systems for new green homes if they don't already understand it. HVAC for high performance homes isn't old school, and your contractor shouldn't be, either.

5

PLUMBING AND WATER HEATING

Water and energy conservation are the overarching (and interrelated) principles for the plumbing and water heating systems in a green home.

Water conservation is the first step. When occupants use less water, they will consume less energy to heat, treat, and, where applicable, pump it. Fortunately, there are simple and affordable ways to conserve water in new homes. Although builders don't control how much water home buyers will use for cleaning, bathing, or brushing their teeth, specific practices and materials can reduce the amount of water needed to operate a home and preserve landscaping.

OUTDOOR WATER USAGE

To limit outdoor water use, landscaping should minimize turf grass, focus on establishing native and drought-tolerant plants, and include water-efficient irrigation systems and controls. Use the home owner manual to inform customers about outdoor water conservation.[44]

The following practices will encourage efficient outdoor water usage:

- **Use *xeriscaping*.** Minimize the amount of water that will be needed by planting drought-tolerant plants and grasses.

- **Collect and reuse rainwater or *gray water*** from indoor sinks and tubs. You can install rain barrels at downspouts to collect rainwater to water gardens or a more elaborate whole-house rainwater catchment system.

- **Cluster plants** with similar watering needs.

- **Do not install irrigation for lawns.** Grass goes dormant, but remains alive, when rainfall is in short supply. Once rainfall returns, most perennial grasses turn green again.

- **Use drip or other micro-irrigation devices** that apply water slowly at or very near the plants' root zone. This will minimize water losses due to evaporation or runoff.

- **Consider omitting landscape irrigation systems.** Hand watering is a valid approach to limit outdoor water use.

- **Mulch around plants** to retain moisture.

- **Install "smart" landscape irrigation controllers** that monitor soil moisture to control watering.

- **Include information in the home owner manual** about outdoor water conservation such as

  - the best time to irrigate—at dusk and early morning—to minimize evaporation losses;

  - how much water is wasted by overwatering;[45]

  - tips on water efficiency from the local cooperative extension service, water utility, or organizations such as the American Water Works Association (AWWA), which presents ideas for conservation at http://www.waterwiser.org; and

  - why to leave car washing to professionals who can recover and treat (and, in some cases, recycle) the effluent.

HARVESTING RAINWATER

Collecting rainwater is an old-fashioned green building technique, but modern systems have made collection and maintenance easier. In some jurisdictions, rainwater can be used for toilet flushing, clothes washing, and for other functions that don't require potable water, in addition to irrigation. Although the financial return for collecting and reusing rainwater is greater in areas where water is more expensive, the environmental benefit of reducing overall water usage is ubiquitous. The American Rainwater Catchment Systems Association (ARCSA) offers training and an accreditation for rainwater harvesting design and installation professionals. Components of a typical rainwater system include outdoor tanks, separate lines for non-potable water, and an indoor pump (figs. 5.1a and b).

FIGURE 5.1A | Rainwater harvesting tanks

These tanks are ready to be buried underground as part of a rainwater harvesting system that serves outdoor water uses, the laundry, and toilets using separate, color-coded plumbing lines. *(Jeannie Leggett Sikora)*

FIGURE 5.1B | Indoor rainwater pump

The indoor pump simplifies troubleshooting and maintenance. *(Jeannie Leggett Sikora)*

INDOOR WATER USAGE

In most homes the biggest water consumers are toilet flushing, clothes washing, and showering. Fortunately, low-flow toilets have improved and their cost has declined. Whereas a *dual-flush* toilet that offers two flushing volumes (typically 0.8 and 1.6 gallons per flush) for the different types of waste cost about $400 four years ago, today you can purchase a similar toilet for less than $125.

The simplest way to select water-efficient toilets, faucets, and showerheads is to look for the WaterSense label. Products bearing the label meet EPA criteria for water consumption. EPA also has an online product directory (http://tinyurl.com/EPAwatersense).

Much of the water used in homes is wasted. It's difficult to pinpoint culprits, but bad habits like allowing water to run while you brush your teeth, or having to run water to get a heated stream are among the common water-wasting practices. Although the former practice requires altering behavior, new technology can alleviate the latter one. If a home delivers hot water quickly to where it's needed, less will be wasted. For example, if a bathroom with high traffic is far from a central water heater, you can install a second tankless water heater at the point of use (fig. 5.3).

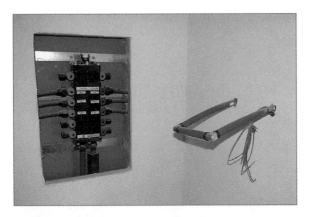

FIGURE 5.3 | Rough-in plumbing for a remote tankless water heater

Solar preheated hot water (run through the line at right) serves a tankless water heater adjacent to a master bathroom that is far from the main water heater. The remotely located heater (left) provides hot water quickly to the bathroom and reduces the amount of water wasted waiting for hot water to arrive at the tap. (*Jeannie Leggett Sikora*)

Just like the ENERGY STAR label, the WaterSense label can be applied to a new home that meets the program's criteria for water conservation. For information about the WaterSense labeling program for new homes, visit http://www.epa.gov/WaterSense.

Hybrid Hot Water

Students at the Lancaster County (Pennsylvania) Career and Technology Center installed a hybrid water heating system in a home they built. The system uses the sun's energy to preheat water and includes two tankless electric backup units—one next to the master bath and another centrally located one to serve the rest of the house. The system delivers hot water almost instantaneously to the remote master bath and cuts water waste.

WATER HEATING ENERGY USE

As the thermal performance of buildings improves and heating and cooling loads diminish, water heating has become a larger fraction of whole-house energy usage. Water heating comprises nearly one-quarter of annual household energy consumption in U.S. homes built in 2000–2005, according to the U.S. DOE.[46] More careful design and selection of water heating systems could reduce this usage significantly.

Hot Water Distribution

Planning how to distribute water throughout a home is the first step to improving hot water system efficiency. When you are designing the home, consider how to minimize distribution pipe. This will reduce water waste while users wait for a hot water stream at the tap. It will also cut the energy waste after faucets are shut off and the energy dissipates from water left standing in the pipes. With shorter distribution pipes, home owners also will enjoy shorter wait times for hot water at the faucet.

To get hot water quickly to where it's needed and, thereby, reduce water waste and water heating energy waste, consider these techniques:

- Cluster hot-water-using appliances and rooms (e.g., baths, laundry, kitchen).

- Stack bathrooms in multistory homes.

- Locate the water heater in a central location and as close to points of use as possible.

- Use the smallest diameter piping allowed by code.

- Use a manifold distribution system rather than a trunk-and-branch system. This will permit smaller diameter piping.

For durability and even more energy savings, follow these best practices:

- Insulate hot water pipes.

- Insulate cold water pipes run through unconditioned areas (to prevent *sweating*).

- Do not run water pipes through exterior walls.

Beyond installing an elegant distribution system, you can also use a circulation pump to move hot water quickly to its destination. Timers, thermostats, motion sensors, or manual switches activate circulation pumps and make hot water available almost instantly at the tap. Although all of these systems will reduce the amount of water wasted while users wait for hot water to reach the faucet, manual switches are the best option for eliminating energy waste (from pumping electricity and from hot water that circulates but is not used).

Heating Water

Americans are accustomed to abundant hot water. This requires a high performance water heating system. Home buyers want a water heating system that delivers hot water quickly, provides ample supply for simultaneous or consecutive uses, and maintains a consistent temperature during a shower (even if a toilet flushes at the same time).

The traditional vehicle for providing these conveniences has been a full hot water tank. As average home size and the number of baths increased, builders responded by installing larger tanks. But maintaining a tank full of hot water is inefficient and sometimes inadequate (hot water supply can run out during or after back-to-back showers). And it doesn't solve the problem of unused water flowing down the drain before the hot water gets to the tap.

WATER HEATER EFFICIENCY

The electric resistance tank water heater has been a popular option for a long time. Electric resistance tank water heaters are practical: they are inexpensive to install, simple to operate, and usually an adequate solution for providing hot water. However, they cost the most to operate and are the least efficient option for providing hot water. Meanwhile, a recent study comparing water heating systems found the operating costs of an average system varied by as much as a factor of three—from $150 per year for a solar water heating system with tankless gas backup to $447 per year for an electric resistance tank.[47]

Energy factor (EF) defines water heater efficiency: it represents the amount of useful hot water produced for every unit of energy input. Although it has historically been a number between 0 and 1, electric heat pump water heaters take advantage of the usefulness of electricity (and the energy in the air surrounding the water heater) to reach an EF of 2 or more. However, EF does not tell the whole story. To compare the efficiency of gas and electric water heaters, you must examine the *source energy*.

Although an electric resistance water heater may have an EF of close to 1, the efficiency with which it produces hot water does not account for the inefficiency of generating and distributing the electricity. Thermal energy is the least versatile type of energy. Unlike electricity, which serves various purposes, thermal energy can perform only one function—heat. This fact raises a question: is it logical to use superior electrical energy to produce inferior thermal energy?

As a rule of thumb, for every unit of electricity that arrives at a house, more than three units of source energy (such as coal or natural gas) were input at the power plant. That is, only 30% of the source energy actually makes it to the home as electricity. For every unit of natural gas supplied to a home, on the other hand, there is less than 10% loss in distribution.[48] Therefore, EF doesn't tell the whole story when comparing the efficiency of heating water with various fuels. Because of the inherent inefficiency of converting electricity directly to heat, an ENERGY STAR label is not available for electric resistance water heaters.

Hot Water Production and Storage

By storing a specific volume of heated water, hot water tanks incur standby energy losses of approximately 10%, depending on usage, tank design, and fuel type. Merely eliminating the tank increases efficiency. However, other factors, including initial cost and capacity, influence whether to choose a tank or a tankless water heating system.

Heating water instantaneously, as in a tankless system, demands substantial of energy. Moreover, although electric tankless water heaters may suffice in warmer climates, they may not have sufficient capacity for homes in colder climates. For example, an electric tankless water heater, heating water from 45°F–105°F for two simultaneous showers (using WaterSense-labeled, 2.0 *gpm* showerheads) will draw approximately 32 kW. This would require the largest electric tankless unit on the residential market. Capacity will increase as the incoming water temperature warms. Gas tankless water heaters with much larger capacities, ample for instantaneous residential hot water production, are available. Because using electrical resistance elements to heat water is inherently inefficient, as previously discussed, and tankless electric heaters may underperform, electric tankless water heaters are best in high performance homes as auxiliary units for solar preheated water or as point-of-use heaters for remote baths.

ENERGY STAR FOR WATER HEATERS

Water heaters that meet the following efficiency criteria (and adhere to specific capacity, warranty, and safety constraints) are eligible to receive the ENERGY STAR label:

- Gas tank: EF ≥0.67

- Gas tankless: EF ≥0.82

- Gas condensing: EF ≥0.80

- Electric Heat Pump Water Heater: EF ≥2.0

- *Solar fraction (SF)* ≥0.5 (the portion of the household's hot water that is anticipated to be supplied by solar energy)

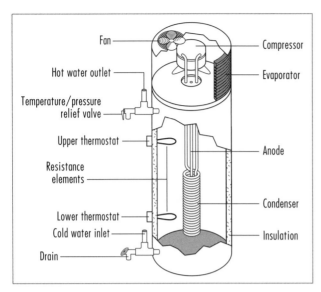

Fan — Compressor

Hot water outlet — Evaporator

Temperature/pressure relief valve

Upper thermostat — Anode

Resistance elements

Lower thermostat — Condenser
Cold water inlet — Insulation

Drain

FIGURE 5.4 | **Heat pump water heater**

Heat pump water heaters consume a little more space but they use about half as much electricity as a conventional electric resistance water heater. (U.S. DOE)

Considering both performance and capacity, using gas to keep a tank of hot water ready is inefficient. When gas tankless technology was new, maintaining precise control over water temperature was a problem, but it is not anymore. Although tankless water heaters often cost more than tank heaters, energy savings may tip the scales towards tankless gas heaters in the long run.

You may prefer electric water heating in your green home to avoid having combustion appliances inside the home or because relative fuel costs make it an economical choice. Except in the warmest climates, heat pump water heaters (the most efficient electric option) usually require a tank for hot water storage to boost capacity. Heat pump water heaters have an added benefit in cooling-dominated climates: they dehumidify and cool the surrounding space (fig. 5.4). They take up slightly more room than a conventional, electric tank water heater, but are approximately twice as efficient. They are an excellent choice for an all-electric home.

Although they rely on hot water storage, solar water heating systems can be efficient and economical when you combine them with a tankless auxiliary heat source. For more information on solar water heating, refer to chapter 8.

Gas water heater variety and efficiency has grown. Choices include condensing units (which extract heat from the flue gases to boost efficiency), non-condensing units, and hybrid systems with small tanks to boost instantaneous capacity. Whereas old natural draft gas water heaters were around 50%–60% efficient, the most efficient systems on the market today achieve 98% efficiency.

Yet another category of water heaters is combination heaters that use a single piece of equipment for space and water heating. Typically, a storage tank is set up as another zone in a hydronic heating system. Water is heated by the furnace or boiler and stored in a separate tank with a priority switch for water heating. Depending on the efficiency of the heating system and the standby energy losses of the tank, these systems can be part of an energy-efficient, economical solution.

Energy Recovery

Hot water flowing down a shower drain contains thermal energy. Some manufacturers have developed heat exchangers to capture and use this energy. There are different types of systems, but those that do not store energy are the simplest and least expensive to install and operate. In these systems, a heat exchanger (typically copper pipe wrapped around a section of vertical drain pipe) captures the energy in the hot water flowing down a drain to preheat the incoming cold water to a shower—these systems work optimally in multifamily buildings and in homes where showers are long and often consecutive. Because the heat exchanger needs time to warm up in order to reach maximum efficiency, it provides less savings during short showers and cold-start (e.g., nonconsecutive) hot water uses. Exchangers that don't store energy do not recover heat from baths.

A de-superheater is another type of energy recovery device. De-superheaters recover waste energy from a ground-source heat pump (GSHP). They enhance cooling efficiency and generate usable hot water from the waste heat. In heating mode, de-superheaters zap a bit of the efficiency from GSHPs but they still provide efficient water heating. De-superheaters are an ideal add-on for GSHPs in cooling-dominated climates. Alternatively, GSHPs can be controlled to provide dedicated water heating. But in that case, the energy of pumping and other factors detract from the efficiency.

DOING MORE WITH LESS

Although water conservation has long been popular in arid climates, many builders and home owners in other climates have not similarly emphasized water conservation. Regardless of its location, a green home should conserve water and hot water energy use.

Historically, residential research and development projects have focused less on domestic hot water energy efficiency than on other areas of residential efficiency. As building envelopes, lighting, and HVAC systems have improved, however, development efforts have shifted toward other energy uses like water heating. High performance systems improve the energy efficiency of water heating while maintaining or improving performance.

Meanwhile, programs like WaterSense makes it easy to find products that will perform the same tasks with less water waste. By adding technology to minimize water waste and informing home owners about habits that conserve water, you can have a tremendous positive impact on the conservation of an essential resource.

6

LIGHTING

Sunlight pouring into windows creates warmth, evokes comfort, and helps sell homes. But as discussed previously, windows (with the exception of those in passive solar homes) also compromise thermal efficiency. On the other hand, well-placed windows reduce the need for artificial lighting so they contribute to energy-efficient lighting.

The amount of energy used to light an American home varies with home owners' habits and a home's latitude. Homes use more energy for lighting when there are fewer daylight hours. The U.S. DOE estimates that lighting accounted for roughly 10% of total household energy usage (and more than 10% of the energy budget) nationally in 2011; that percent was trending slowly downward.[49]

As with water conservation, builders can't control home owner behaviors that waste energy, such as leaving lights on in an unoccupied room. However, they can influence the energy efficiency of residential lighting through design, fixture selection, and controls.

Apply these strategies to create pleasing and efficient lighting:

- Use daylight to reduce the need for artificial light.
- Provide task, ambient, and accent lighting in each room.
- Install energy-efficient fixtures and lamps.
- Add controls such as multiple switches, motion detectors, timers, photo sensors, and dimming switches.
- Inform home owners about best lighting practices and technology.
- Consider installing energy-monitoring devices that offer real-time feedback on energy consumption.

DESIGN

Simple design concepts like *daylighting, layering,* installing multiple types of controls, and adding automation can save energy.

Daylighting

Daylighting is dynamic and appealing to home owners, but as you incorporate natural light in a home design, you must also control heat, glare, and UV degradation of building materials. Daylighting strategies include the following:

- Strategically placing windows on all sides of a building, including the southern exposure, while using overhangs to prevent overheating and glare in the summer.

- Using skylights with shades or tubular skylights, which are more thermally efficient than conventional skylights.

LIGHTING DESIGN CONCEPTS

Use layered lighting design so home owners have only the light they need—brighter light for specific tasks and lower light for general lighting of rooms. General lighting includes ceiling fixtures. Task lighting includes desk and kitchen counter lights. Accent lighting provides architectural interest. When you incorporate all three types of lighting and provide separate switches for each one, home owners can control lighting for comfort and efficiency. For example, you can provide under-cabinet lighting for cooking and other lighting at the kitchen table. Lower-wattage task lighting at these locations may eliminate the need to turn on higher wattage general lighting.

LIGHTING DESIGN RESOURCES

The Lighting Research Center (http://www.lrc.rpi.edu) offers online lighting advice and room-by-room designs in two resources: *The Builders Guide to Home Lighting*[50] and the *Lighting Pattern Book for Homes.*[51] *The Lighting Pattern Book for Homes* offers room-by-room plans for energy-efficient lighting. *The Builders Guide to Home Lighting* offers advice on selection and installation of energy-efficient lighting. Search for the titles on the home page.

SELECTING FIXTURES AND LAMPS

Lighting technology has advanced so we can enjoy efficiency without the bluish hue and flickering of old fluorescents. Lamps (aka, bulbs) are available that produce the same warm light we are accustomed to from incandescent bulbs but they use a fraction of the energy. Fluorescents and LEDs produce more *lumens per watt (LPW)* than incandescent bulbs, which translates to the same light output for less energy input.

New lighting labels simplify fluorescent and LED lighting selection. Look for *color temperature (CT)* of 2700K–2800K and *color rendering index (CRI)* >80 to best approximate incandescent lighting.

Besides efficiency, fluorescent and LED lighting have two other benefits:

1. They last longer than incandescents. This makes them great for hard-to-access lighting because lamps don't need to be changed as frequently.

2. They generate less heat so they reduce the cooling load in cooling-dominated climates.[52]

LED lights are directional, so, historically, they have been most practical for task lighting. However, with advancing technology, they are increasingly being used for general illumination. LEDs are well suited for desks, stairs, as downlights, and for under-cabinet kitchen and landscape lighting.

Compact fluorescent lamps (CFLs) are ubiquitous. You can find them in a variety of shapes, wattages, and color options everywhere other lightbulbs are sold. Dedicated fluorescent and LED lighting fixtures also are widely available.

The ENERGY STAR Advanced Lighting Package applies to homes in which at least 60% of the light fixtures (and all ceiling fans) are ENERGY STAR labeled. Many green builders are upping the ante by installing 100% high efficiency lighting.

ENERGY STAR labels more than 32,000 fluorescent light fixtures and more than 1,000 LED light fixtures, not to mention lamps and ceiling fans. The array of styles is a boon to builders who want to install dedicated high-efficiency fixtures. Your local lighting supply store

The American Lighting Association (http://www.americanlightingassoc.com) offers online videos of modern lighting designs and options.

or manufacturers' catalogs may be the best source of options for energy-efficient light fixtures and ceiling fans. Despite the operational cost savings for the home owner, dedicated fixtures tend to cost more than conventional ones so you may need to adjust the lighting budget for this new paradigm.

Beyond the fixture and lamp, other factors impact the energy efficiency of lighting. For example, recessed lights that penetrate the building envelope can contribute to air leakage and condensation. Therefore, avoid placing recessed lights in the top floor of homes. If you choose to place them there, they should be airtight, *IC-rated*, and sealed to the drywall below.

LAMP LABELING CHANGES

Federal labeling requirements for lamps replace information about wattage with lumen output. Lumen output labeling fosters comparisons of energy-efficient lamps, which produce more lumens per watt (LPW) than their less-efficient counterparts. The new packaging also will include information about *CT* and *CRI* so you can select bulbs that produce warm, inviting light. For light most similar to incandescent quality, select lamps with 2700K–2800K CT and CRI >80.

You can use recycled-content light fixtures (such as those by Eleek, of Oregon) and repurposed light fixtures. However, recycled-content and repurposed fixtures are not as widely available and tend to cost more than conventional energy-efficient fixtures.

CHOOSING CONTROLS

Builders can help reduce energy needs for lighting by installing controls and providing information to home owners about how to operate lighting most efficiently. Controls such as dimmers, occupancy/presence sensors, motion sensors, and *photocells* can reduce the effect of home owner habits that waste electricity.

Dimmer switches and timers allow home owners to create ambience and save energy. Although you can use dimmers with incandescent bulbs, energy use does not decline proportionally to the light levels. In fact, dimming incandescent lamps reduces their efficiency in lumens per watt. Therefore, use efficient CFL or LED lamps on dimmer switches, but keep in mind that many screw-type CFLs (which fit conventional fixtures) are not dimmable.

When you install dimmable CFLs, match the dimmer switch to the CFL by using the information included on the product's packaging. For hard-wired fluorescent lamps that have the ballast integrated with the fixture, match the dimmer switch to the ballast. Your lighting distributor or the manufacturer will know which switches are compatible. For LEDs, dimming is more complex so consult the manufacturer for a list of compatible dimmer switches. When using any electronic lighting controls, such as timers, select controls that are rated for the controlled circuit's wattage.

IMPROVING LIGHTING EFFICIENCY

The following strategies will help improve lighting efficiency:

- Add natural lighting without significant thermal energy implications by using transom windows, tubular skylights, or operable (north-facing) traditional skylights.
- Limit the number of lamps operated by a single switch.
- Install task lights to reduce ambient lighting needs.
- Install hard-wired fluorescent or LED fixtures, especially in areas where lights are typically on for long periods—kitchen, dining room, living room, utility room, and outdoors. These dedicated fixtures do not support screw-in lamps so home owners will not be able to replace them with inefficient incandescent bulbs.
- To control air leakage, avoid recessed lights in upper-floor ceilings.
- Consider tubular skylights where natural lighting is desirable, such as in dressing areas.
- Add lighting controls such as dimmer switches, photocells, and occupancy sensors to reduce unnecessary illumination.

- Install screw-in compact fluorescent (or LED) lamps in conventional light fixtures.

- Select lamps that can be dimmed and match proper dimming controls to them.

Photocells, such as those in nightlights that illuminate only at dark, detect light and activate lighting when levels fall below a threshold. When using photocells for outdoor lighting controls, ensure that the fixture also includes a motion detector so outdoor lighting does not operate all night. Photocells, dimmers, and shading control also can be part of a daylighting strategy that automatically adjusts lamp output (for the level of natural light available) and shading (to optimize solar gain). Although these high-tech controls are more prevalent in commercial buildings, they are available and adaptable for residential use. They can be integrated with HVAC and security systems for a fully automated green home (check out programs like Crestron's Green Light at http://www.crestron.com/products/green_light_residential_lighting_control/.)

Beyond simply reducing energy usage, controls can help reduce energy costs when they are integrated into a smart utility grid solution that receives information from the utility (through a smart meter) and automatically controls loads to use electricity when it is cheapest or most abundant.

There are two types of occupancy/motion sensor technologies: *passive infrared (PIR)* and *ultrasonic*.

PIR technology senses body heat from multiple zones within its detection area. When it detects a new heat source in a zone, it registers the zone as occupied. PIR sensors require a clear path around the room and need a relatively large amount of motion (about the magnitude of a person taking a step) to function. Because they must detect motion in order to keep a light on, they make more sense in kitchens and hallways than in rooms where people won't be moving around as much.

Ultrasonic sensors are active sensors that emit sound waves (outside of the range of human hearing). Based on the time it takes for the wave to return, the sensors detect even slight motion. Ultrasonic sensors do not need a clear path of detection so they are suitable for offices, living rooms, and possibly bathrooms. Although occupancy sensors are popular in commercial buildings, residential designers and builders have not adopted them widely.

If you want to use occupancy sensors, install products that also include photocells so the system won't activate artificial lighting when there is sufficient daylight. Under certain circumstances, occupancy sensors can be convenient and efficient but they may also reinforce the bad habit of not turning off the lights!

SKYLIGHTS

Although they are technically part of a daylighting strategy, skylights are discussed separately here because of the additional design and operation considerations to ensure the skylights function as an integral part of a high performance home.

SKYLIGHTS AND U-FACTOR

Because skylights are usually installed on a sloped roof and project beyond its plane, they are laboratory tested for the NFRC label in a sloped, projected configuration. This testing configuration (which promotes thermal currents to develop between the panes), causes U-factor to be higher for double-pane skylights in a sloped configuration than it would be for a vertical installation. Depending on the installed slope, actual U-factor can differ from the NFRC-rated value.

Skylights have a much lower insulating value than an uninterrupted insulated ceiling, so using them garners an energy penalty. You can offset the penalty by incorporating the following design principles when you specify skylights:

- Place skylights precisely where light is needed.
- Add shading and operating controls when the budget allows.
- Place operable skylights high in the space to encourage natural ventilation via the chimney effect.
- Consider using tubular skylights for interior spaces like walk-in closets, hallways, and bathrooms. They have a much smaller square footage than conventional skylights and, hence, a lower energy penalty.

When properly selected, skylights can add natural lighting to a house and reduce the total amount of glass in the building. Recent cold-climate studies in Europe suggest that this strategy may eliminate the energy penalty associated with skylights.[53] In fact, light transmitted through a sloped skylight is about 3–10 times more intense than light on a vertical surface. Of course, windows are not just used for lighting, they also provide a view or a connection to the outdoors. However, where your home owners want privacy and natural lighting, skylights can be a smart choice.

In cold climates, condensation will probably form on skylights as warm air rises and hits the cool glass surface, especially in humid areas like bathrooms and kitchens. Some manufacturers have incorporated features to handle the condensation so that it will not damage building materials, reduce the skylight's durability, or both. When you research products, ask how they handle condensation. In warm climates, include shading devices on skylights so the sun doesn't overheat a space, especially a sleeping area.

Skylights and installation methods have come a long way. At least one manufacturer offers a 10-year warranty on installation. Always select an experienced skylight installer who was trained by the manufacturer on methods and compatible materials. Most manufacturers back their product as long as installation instructions are followed precisely.

AMBIENCE AND EFFICIENCY

Lighting can introduce a "wow" factor into a new home. Choices for stylish and efficient lighting abound. Use the resources available through programs like ENERGY STAR and the Lighting Research Center for guidance on creating ambience and functionality that customers will enjoy. By strategically planning lighting and using off-the-shelf fixtures, you can create a highly efficient and attractively lit home and landscape at a reasonable cost.

7

APPLIANCES AND PLUG LOADS

Appliance efficiency has improved dramatically in the last few decades. At the same time, electricity use for consumer electronics has skyrocketed. Although finding energy-efficient appliances is easy, expecting home owners to unplug their communication and entertainment devices would be a more difficult proposition.

APPLIANCES

Appliance choices have a long-term impact on home water and energy usage. And since appliances use an expensive energy source (electricity), efficiency savings can net significant cost savings.

The most energy-efficient clothes and dishwashers achieve high efficiency by dramatically cutting water use per cycle. In other words, the energy efficiency of those appliances is directly related to the amount of hot water they use.

Clothes Washers

Today's high-efficiency clothes washers use about half the water of older models.

The ENERGY STAR program rates clothes washers according to the following criteria:

- **Modified Energy Factor (MEF)** considers how much energy the machine uses, the estimated energy use to heat the machine's water, and the estimated energy use to dry the clothes after washing. The lower the MEF, the more efficient the machine is.

- **Water Factor (WF)** considers how efficiently the machine uses water. The lower the WF, the more efficiently the machine uses water. WF levels the playing field to compare washers of various capacities, unlike estimated *kilowatt*-hours (kWh) used per year.

Compared with ENERGY STAR criteria, the yellow EnergyGuide label is more difficult to compare across different types of washers and capacities. Therefore, when you want to select the most energy-efficient clothes washer, refer to the ENERGY STAR label and product information regarding MEF and WF.

Although high-efficiency top-loading washers are available, front-loading machines achieve the highest ratings for energy and water efficiency (table 7.1).

TABLE 7.1 | Standards of efficiency and range of efficiency for ENERGY STAR-labeled clothes washers on the market as of 2011*

	Water Factor	Modified Energy Factor
Federal minimum criteria	≤9.5	≥1.26
ENERGY STAR criteria	≤6	≥2
Top-loading machines	2.9–6.0	2–2.6
Front-loading machines	2.7–5.8	2–3.4

*Based on ENERGY STAR list of qualifying appliances

Dishwashers

Almost all dishwashers on the market today bear an ENERGY STAR label. Although the government mandates that new dishwashers use 355 kWh or less per year (including estimated electric water heating energy), some ENERGY STAR models cut electricity use by about half (the most efficient models tend to also cost the most). Of the ENERGY STAR-labeled dishwashers on the market, the most efficient models use about one-fourth the water of the least efficient models.[54]

Refrigerators

Because of increasingly stringent federal mandates on refrigerator efficiency, today's refrigerators use only a fraction of the energy of pre-1992 refrigerators. ENERGY STAR models consume less than 2 kWh per day with some full-size models consuming as little as 1 kWh per day. The most efficient models have the freezer on top and no through-the-door ice and water dispensers. Yet, because refrigerators are no longer the energy hogs that they once were, convenience, capacity, interior design, and aesthetic appeal may influence refrigerator selection as much or more than the energy consumption numbers.

Disposals

Garbage disposals are ubiquitous in new homes served by municipal water systems. There are no efficiency ratings for them because they use relatively little electricity. Therefore, selecting for energy efficiency is not straightforward. The most energy-efficient option for the home owner, of course, is to not use a disposal. Omitting the disposal both eases the demand on water treatment plants and saves water because without the disposal you don't need a stream of water to wash food down the drain. However, an LCA study of food waste treatment options concluded that filling landfills with food waste creates more potential for global warming than energy usage from garbage disposals does.[55]

Builders can encourage home owners to compost, as an alternative, by providing an area in the kitchen for collecting food waste. There are commercially available under-counter mounted and in-cabinet mounted bins. Planning space for collecting kitchen scraps can be part of your overall strategy for maintaining the "greenness" of a new home after the sale.

Dehumidification systems

Controlled humidity enhances comfort in the home, makes it less prone to moisture problems, and increases its durability. A whole-house dehumidification system (or an AC unit capable of stand-alone dehumidification) can be integrated into HVAC design in a high performance home. Stand-alone dehumidifiers can manage high relative humidity that can plague consistently cool basements and crawl spaces. Because the cooler basement temperature and higher relative humidity go hand in hand, even a basement with impeccable foundation drainage could benefit from a dehumidifier. To promote humidity control, you can include information about stand-alone dehumidifiers in your home owner's manual.[56]

PLUG LOADS

In addition to large appliances, home owners plug many consumer electronics into electrical outlets. These devices, known as *plug loads*, are on the rise despite gains in residential efficiency in almost all other energy-use categories. Plug loads are nearly one-third of the average U.S. home's electricity consumption—on par with electricity use for space conditioning and more than twice the electricity used for lighting (fig. 7.1).[57] According to the U.S. Energy Information Administration (EIA), electricity consumption from plug loads has nearly doubled in the past three decades even as total household energy use declined by a third.[58] One study showed that the average California home spends $150 per year on plug loads and more than half of that is for entertainment.[59] Increasing the efficiency of plug loads could reduce energy consumption in all homes, not just green homes.

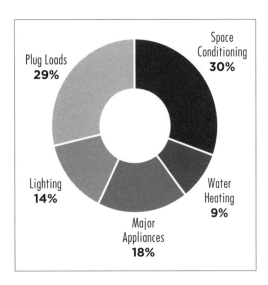

FIGURE 7.1 | **Electricity use in U.S. households in 2010**

(U.S. Energy Information Administration, 2012 Annual Energy Outlook)

Builders don't control the number of electronic devices that home owners plug in, but they can help home owners control these *miscellaneous electric loads (MELs)*. For example, to reduce energy use associated with MELs, you can wire circuits so entertainment centers can be shut down with a single switch to prevent the remote-controlled devices from drawing energy while in *standby mode*. You can also provide charging stations where users can deactivate the electricity supply with the flick of a switch (to prevent transformers from drawing power after a device is fully charged). But the most successful conservation method may reside in helping home owners understand how they are using energy.

A Little Information Can Go a Long Way

When consumers understand how they use energy, they tend to take steps to reduce their energy consumption. Studies have shown that when they get real-time information about energy consumption, households reduce their energy use by 5% to 15%.[60] Utility smart meters and off-the-shelf energy monitoring and control systems help home owners understand energy usage and manage energy consumption. Many systems now allow home owners to manage their home's energy usage remotely via the Internet. Some systems provide whole-house control of lighting, security, and entertainment.

The following strategies can influence home owners' energy usage after they move in:

- Include compelling information about MELs in your home owner manual.

- Inform home owners about ENERGY STAR-labeled consumer electronics and their impact on standby power usage (they use less of it than non-labeled devices).

- Install devices that indicate where household energy is being consumed in real time.

- Provide a whole-house energy management system to monitor and control electronics. These systems can be installed with dedicated wiring, but there are also many wireless systems on the market.

- Wire entertainment area circuits so a single switch can shut off all devices.

- Provide a "charging" station area with a switchable power outlet to prevent devices from drawing power when they are not charging.

THE POWER OF THE PLUG

Homes today use about the same amount of energy as they did in 1978. Yet, over that time, efficiency has improved across nearly all end uses of residential energy. As average home size increased by a third, heating energy use declined by more than a third. Major appliance efficiency has reached an all-time high. At the same time, the growing number of consumer electronics in homes has contributed to almost doubling appliance and consumer electronic energy use.[61] Cutting-edge green homes can provide solutions that will mitigate this upward trend.

8

GENERATING RENEWABLE ENERGY

Energy efficiency is often a wiser investment than a renewable energy system. It is frequently less expensive to save a kWh by making a home more efficient than to generate a kWh with a renewable energy system. There are exceptions to this rule of thumb, of course, depending on local climate and utility rates. For example, there is often a quick payback from using a solar thermal water heater instead of an electric tank. Financial incentives, as well, can tip the scales toward renewable energy systems.

Although they might not have the highest return on investment, renewable power systems are popular—thanks in part to widespread utility, state, and federal government incentives. Whether they are selected for insurance against future energy price hikes, to make a statement, to offset a home's energy consumption, or for any other reason, residential renewable energy systems are here to stay. Aesthetics have improved since the 1970s solar movement when panels were often mounted on ugly "spider legs." Panels are now mounted flush with the roofline (even if this sacrifices a bit of efficiency) by third-party accredited installers.

The best place to go for comprehensive information about federal, state, local, and utility incentives, such as tax credits, loans, or grants for renewable energy systems is the Database of State Incentives for Renewable Energy (http://www.dsireusa.org).

SOLAR THERMAL ENERGY

Solar thermal energy is easy to collect—just imagine walking barefoot on an asphalt driveway in the summer—and there are many options for installing a solar thermal system. A solar thermal system can collect energy to heat air or water. Since water can store much more energy than air can, solar thermal water heating systems tend to be more popular than air heating systems. Most green builders use passive solar design, rather than an active solar thermal system, to heat indoor air.

Solar thermal systems offer the quickest payback when they replace electric resistance water heating. In an all-electric home, solar thermal collectors and an electric tankless auxiliary unit can be a great high performance water heating option.

Four components typically comprise a solar thermal system:

1. A roof-mounted absorber that collects thermal energy
2. A tank for storing solar-heated water
3. A pump with controls that activate the system when there is sufficient solar energy
4. An auxiliary energy source for water heating when there is not enough solar energy to meet household needs

In most systems, the storage tank includes an internal heat exchanger. The fluid that absorbs the solar energy (a glycol-water mix in climates that experience freezing temperatures) heats the potable tank water. Instead of using pumps, some systems rely on *thermosiphoning*, the natural buoyancy of hot water, to move the fluid through the collectors. Collectors can be *flat-plate* or *evacuated tube*. Evacuated tube collectors are lighter and are partially assembled on-site so they are somewhat easier to install (fig. 8.1). They also offer slightly higher efficiency at very cold temperatures but approximately equivalent annual efficiency as flat-plate collectors.

For very mild climates, you can opt for *integrated-collector storage (ICS) systems*. These systems combine collector and storage into one roof-mounted unit, instead of having a separate absorber and storage tank. These systems work best in sunny regions with moderate temperatures.

Almost all roof-mounted systems have integrated flashing, mount flush with the roof, and offer extended warranties of at least 10 years.

If you don't wish to incur the expense of a solar thermal system but want to encourage its future use, consider roughing-in the plumbing for a future solar thermal installation. A rough-in includes supply and return piping, or a conduit for the piping, which runs from the

CHOOSING A SOLAR WATER HEATING SYSTEM

Don't be overwhelmed by the array of solar water heating systems and options for configuring them. The Solar Rating and Certification Corporation offers a directory of *OG-300 certified systems* and there are experienced and certified solar thermal installers. The North American Board of Certified Energy Practitioners (http://www.nabcep.org) can help you find certified professionals in your area. Choosing a system with an OG-300 certification earns points in most green building programs and ensures system components will function together.

An OG-300 rating incorporates two measures: *solar energy factor (SEF)*, a performance rating for solar water heating systems, and *solar fraction (SF)*, the fraction of the energy delivered that is expected to be provided by the sun rather than by an auxiliary energy source. SEF is not equivalent to SF, but you can convert it. ENERGY STAR labeling uses SF to rate solar water heating systems and requires a rating of 0.5 or greater. ENERGY STAR also rates durability and other factors.

mechanical room to the roof. Ideally, future panels will be mounted on a south-facing roof at an angle approximately equal to the latitude. The roof should be free from shading most of the day. Panels will extend over 50–100 sq. ft. of clear roof area.

PHOTOVOLTAIC SYSTEMS

With financial incentives, rising electricity costs, and declining prices for solar electric systems, rooftop *photovoltaic (PV)* systems seem to be sprouting up everywhere. People tend to prefer a PV system over solar thermal systems, possibly because the benefits are more tangible: you can see a meter showing electricity production. In addition, PV systems have no moving parts so they are relatively durable and easy to maintain.

FIGURE 8.1 | Evacuated tube solar collector

Evacuated tube solar collectors are better than flat-plate solar collectors at preventing heat loss so they are more efficient in cold weather, but the two types of collectors produce approximately equivalent annual thermal energy. *(Jeannie Leggett Sikora)*

Most modern PV systems connect to the utility grid. When the sun isn't shining, the power company supplies electricity and when the PV system produces more power than the home is using, the system sends the excess electricity to the utility company. The home owners often get credit for the full retail value of the electricity their system sends to the power company.

Unlike off-the-grid power systems that rely on battery storage to power the house when the sun sets, systems connected to utility power have optional battery storage, continuous power supplies, or both. These options provide electric power only to important loads—such as sump pumps, refrigerators, medical and communication devices, and egress path lighting—during power outages. Because battery storage adds cost, many grid-connected PV systems simply go offline when the utility power goes out. Figure 8.2 shows a typical roof-mounted PV system.

FIGURE 8.2 | PV on Roof

PV panels adorn the south-facing rooftop on the front of this Pennsylvania home. *(Jeannie Leggett Sikora)*

All U.S. public utilities must provide net metering, which measures excess electricity sent back to the utility (by having the meter spin backward or through separate metering) and credits the home owner. State laws and utilities' policies vary on whether credits carry over to future billing cycles, the price paid for excess electricity, and connection fees. Refer to http://www.dsireusa.org for specific regulations.

A typical PV system includes an array of panels, an inverter which transforms DC power to AC power, utility interconnection hardware, and switches and safety controls. PV panels generate *direct current (DC)* electricity which runs through an inverter to create *alternating current (AC)* for use in the home. System components must match to provide optimal output and minimal efficiency loss. Consult with your PV installer on pre-engineered system options. Other factors that affect system efficiency include orientation and tilt angle of the panels, obstructions such as shadows (even small shadows can dramatically reduce output), dust, dirt, snow, and temperature (efficiency declines as panel temperature rises).

You can mount PV panels on roofs, on shade structures like trellises, or on ground racks, or you can use *Building Integrated Photovoltaics (BIPVs)*. A series of panels tied together is called an *array*. Its size, measured in kilowatts (kW), is the theoretical instantaneous maximum power output at full sun. Many factors (dust, heat, wiring losses, and inverter losses) impact instantaneous output, so annual output predictions that consider local weather data and precise system design features can be a better indicator of a system's capability.

When you are planning a PV system, consider these factors:

- Local interconnection rules (get the utility involved as early in the process as possible)

- Building and electrical codes governing PV systems

- Where to locate the array (the optimal location is on a south-facing slope free from shading from other buildings, adjacent roof area, roof penetrations, and trees)

- System options and pricing

- Added roof load from the PV panels

You will need to specify the capacity criteria for the system: do you want the system to meet 100% of annual electricity needs to create a "net-zero" energy home, or do you simply want to shave monthly electric bills by 10%? Other factors to consider include the need for backup power in electrical outages, and budget. Section 690 of the National Electric Code addresses electrical regulations for PV systems. Qualified professionals should handle final design and installation. Go to the North American Board of Certified Energy Practitioners directory (http://www.nabcep.org) to find a certified PV system installer.

Some power companies and cooperatives encourage PV system interconnection by streamlining the process but others make interconnection difficult. Know your utilities' requirements and involve them at the beginning to avoid problems later on.

Most PV systems produce 5–10 watts per sq. ft. of array area. Therefore, a 4 kW system designed to produce about half the annual energy for an efficient house in a cold climate requires 400–800 sq. ft. of clear roof area for the array, plus up to 20% more for access around the panels.

A California consulting company has developed a customizable scope of work (http://tinyurl.com/PVscope) you can use in overseeing PV system installation. Expect a typical 2-kW grid-connected system to take 2–4 worker days to install.

WIND POWER

Small wind-power systems can be an economical way to produce on-site power, depending on where they are located and the height of surrounding structures or trees that can reduce wind. A grid-connected wind turbine can be practical for homes where the average annual wind speed is about 10 mph or greater, there are no covenants or codes restricting its use, electricity costs are $0.10 per kWh or higher, and the cost of connecting to the utility is not prohibitively expensive or complex.[62]

Although wind turbines make noise, the sound at ground level is generally less than the noise an AC compressor generates.[63] You must protect the turbines from lightning because of their height.

Small Wind Electric Systems: A U.S. Consumer's Guide covers the basics of wind-power systems for homes. Many states have customized the document with state-specific wind maps and information. Search the Internet for the title but replace "U.S." with your state's name.

The Small Wind Certification Council (http://www.smallwindcertification.org) independently certifies small wind-power systems so consumers can compare products on calculated total annual energy output, sound level, and the rated power of the system at a given wind speed. Although only two systems were certified by the end of 2011, some two dozen systems are pending certification.

HYDROPOWER

If your building site includes a running stream, you may want to consider adding a micro-hydropower system to produce electricity. These systems divert water from the stream to power small turbines that generate electricity. They create little environmental disturbance, and a 10-kW system could provide sufficient energy for a large household, according to the U.S. DOE.

If local codes and water rights allow it, micro-hydro power can provide a relatively consistent electricity supply. The following government agencies may have authority over permits and water rights: county engineer, state energy office, the Federal Energy Regulatory Commission's Small/Low Impact Hydropower Program, and the U.S. Army Corps of Engineers.

Micro-hydropower systems may be regulated under Section 404 of the Clean Water Act. Ask your regional Army Corps of Engineers district office whether you need a permit. The office will guide you through the process. To find your local Corps district office, go to http://www.usace.army.mil/CECW/Pages/cecwo_reg.aspx and select your state in the box at the top of the page.

Like PV and wind-power systems, micro-hydro systems can either stand alone or connect to the power grid.

Use the following equation from the U.S. DOE to estimate how much power you could harness from a stream:

$$\text{power output (watts)} = \text{vertical drop (ft.)} \times \text{flow (gpm)} \div 10$$

RENEWABLE ENERGY AS A SHOWPIECE

A home owner can find tremendous satisfaction and pride in a home that produces its own energy. Renewable energy systems are more visible than efficiency measures and tend to generate more conversation among consumers with an energy-saving mind-set. Although the price of renewable energy generation is falling with technology advances and increasing economies of scale, its cost effectiveness still varies greatly with locality. Natural resource and contractor availability, state incentives, and local market conditions all influence availability. Regardless of cost-effectiveness, renewable energy systems can be an appealing centerpiece of a new green home.

9

PERFORMANCE TESTING, DIAGNOSTICS, AND BUILDING COMMISSIONING

One of the most substantial changes in home building of the past decade is the increase in performance testing for homes. This testing corresponds with a general trend of proactively verifying and quantifying construction quality before a home is occupied, rather than trying to fix problems during the warranty period. The trend is a positive development for builders who apply building science principles in constructing new homes.

Performance testing, diagnostics, and start-up procedures fall under the umbrella of building commissioning—the systematic verification of the operation of each system in a building to meet norms established by various industry organizations. Each organization has its own area of expertise. Long a standard procedure for commercial buildings, commissioning has finally made inroads in residential construction. It changes how builders can talk to customers about home performance. Just as pressure testing for plumbing pipe verifies that the system is leak free, commissioning can help verify that the house will perform as it was designed to.

According to the Energy Performance of Buildings group at the U.S. DOE's Lawrence Berkeley Laboratory, commissioning can address seven key performance areas:[64]

1. Building envelope
2. Cooling equipment and heat pumps
3. Distribution systems (air and hydronic)
4. Combustion appliances
5. IAQ

Best practices for residential HVAC systems such as using ACCA Manual D to develop room-by-room design airflow rates, testing and balancing the airflow to each room, and documenting HVAC system start-up, exemplify building commissioning procedures that started with commercial and industrial buildings.

6. Controls

7. Other appliances

Commissioning tests a home's major systems, measures their performance according to industry-accepted norms, and makes necessary adjustments. Different specialty organizations set the norms. For example, ASHRAE sets standards for air-conditioning systems and ENERGY STAR sets standards for installing insulation. Green building rating and certification programs usually mandate some aspects of building commissioning and offer extra points for others. Because building commissioning is one of the most powerful tools a builder can use to prevent costly warranty problems, and many aspects of building commissioning (e.g., documenting start-up procedures) are low cost, you might as well incorporate a comprehensive building commissioning process.

For a typical high performance home, you could include the following measures as part of the building commissioning process:

- Visual inspection of installed systems
  - Air-sealing measures
  - Thermal and moisture-protection systems
  - Continuity of the air barrier
  - Duct installation and sealing
- Physical testing
  - Whole-house air leakage
  - Duct leakage (total and to the outdoors)
  - Room-by-room air delivery flow rates
  - Whole-house ventilation system air flow
  - Ventilation rate through spot ventilation
  - Pressure differentials across doorways
- HVAC system testing
  - Refrigerant charge
  - Air temperature across refrigerant coil
- Safety testing
  - Combustion safety
  - Radon

BUILDING COMMISSIONING-RELATED STANDARDS AND RESOURCES

Each program and organization sets its own standards for building commissioning. Some are stricter than others. You can ask Home Energy Raters and independent green building program verifiers what is considered high performance in your region. Following is a partial list of building commissioning resources:

- ENERGY STAR Thermal Enclosure System Rater Checklist (version 3 is the latest)

- ENERGY STAR Water Management System Builder Checklist

- ENERGY STAR HVAC System Quality Installation Contractor Checklist (aligns with requirements in ANSI/ASHRAE 62.2 *Ventilation and Acceptable Indoor Air Quality in Low-Rise Residential Buildings* and ANSI/ACCA 5 QI – *2010 HVAC Quality Installation Specification*)

- DOE Challenge Home Quality Criteria (see Chapter 11)

- ANSI/ACCA 9 QIVP: *HVAC Quality Installation Verification Protocols*

- Manufacturer's installation instructions and start-up procedures

- Specifications provided by green building programs
  - National Green Building Standard (ICC-700)
  - LEED for Homes
  - EarthCraft Homes
  - Earth Advantage New Homes
  - Canada's R-2000 Standard (available for free download)
  - Local green building programs

HOW TIGHT IS AN ENERGY-EFFICIENT HOUSE?

A house could theoretically be constructed without any unintentional air leakage if it had an energy recovery ventilation system that introduced and distributed sufficient fresh air. But in practice, tight homes still leak a little. Most U.S. green building and energy efficiency programs offer points for meeting air tightness thresholds. Canada's R-2000 certification program specifies that whole-house leakage must be less than 1.5 ACH50. The NGBS (ICC700-2008) awards the highest possible points to homes that test at 1.4 ACH50 or lower.

The North American Technician Excellence (NATE) HVAC Efficiency Analyst is a certification for technicians who demonstrate comprehensive knowledge of various energy efficiency subjects. The HVAC Efficiency Analyst certification, arguably the most comprehensive NATE certification offered, tests knowledge of relevant building science from design to testing and balancing airflow. Only a small percentage of NATE-certified technicians have accomplished advanced certification but they would be a valuable resource for your green building projects.

QUALITY CHECKS

Would you buy a car if its various systems—lights, motor, steering, brakes, and radio—had not been tested? Of course not. But rarely are all home systems—often installed by discrete trade contractors— systematically tested and operated before a house is sold. Building commissioning provides a systematic methodology to assure that systems were installed properly and functioning as they were designed to so that you, and your customers, can know what level of performance to expect.

10

FINISHING TOUCHES

Now for the fun part—the finishes. Finishes are the elements of a green home that home owners proudly show off, enthusiastically discuss with friends, and which serve as the centerpiece of green home design. The nuts and bolts of the project contribute to performance and staying power; the finishes provide the embellishment that home buyers yearn for.

Paint, flooring, trim, cabinetry, countertops, and plumbing fixtures are among the features that collectively offer abundant choices of green finishes. To keep the selection process manageable, consider overall project objectives first. Although one product may contain more recycled content than another, the more attractive and less expensive product may be produced locally and have less embodied energy. Just as two different books or meals can be equally appealing, two different green products may be just that—different. Yet they may be equally attractive and suitable for a green home. Budget, aesthetics, consumer preference, and product availability often play as much of a role as environmental performance in product decisions—and that is OK. For example, some green builders disdain vinyl siding because of its manufacturing by-products. Other green builders point to its positive attributes: it is durable, requires little maintenance, doesn't have to be painted, and, as an exterior product, won't harm IAQ.

Here are some questions to ask in selecting finishes for green homes:

- How much energy is used in manufacturing the product?
- Are the raw materials natural or man-made?
- Is the product made from postconsumer waste? Postindustrial waste? Virgin materials?
- Will the material off-gas potentially noxious fumes after you install it?
- Is it easy to clean with mild cleaning products?

- How long will it last?
- Can you recycle the product?

The inventory of green products is continually changing so this book does not attempt to catalog them comprehensively. However, tables 10.1–10.5 list the general environmental benefits and concerns within these categories of finishing options: siding and trim, roofing, flooring, countertops, and landscaping. The tables purposefully exclude cost and installation factors because most environmentally preferable systems cost more than conventional ones and some are more difficult to install. Depending on your priorities for the project, and your client's, particular finishes may rise to the top in making selections.

PRODUCT SELECTION TOOLS

You can learn more about green product options from the NAHB Research Center's list of green-approved products (http://tinyurl.com/greenapproved) and the other green resources at http://www.nahbrc.org. Other sources of product information include the International Builders' Show and green building conferences, *Builder* magazine, manufacturers' technical information, and directories and databases such as GreenSpec (http:// www.buildinggreen.com/menus), Oikos (http:// oikos.com), GreenGuard certified (http://www.greenguard.org), and SCS Certified Products (http://www.scscertified.com/products).

Most major manufacturers encourage builders to specify their products for green homes by explaining on their websites how their products qualify for points under LEED for Homes and NGBS. This product-specific information can help you determine if the product is compatible with the project's environmental goals.

In addition, LCA software can help you rank factors such as IAQ, embodied energy, and human toxicity.[65]

SIDING

Because water eventually gets behind siding, the groundwork for high performance siding is laid when the home's weather barrier and flashing system are designed and installed. Yet, there are other environmental factors to consider in deciding which siding product best suits a green home (table 10.1).

TABLE 10.1 | Green properties of siding and trim alternatives

Material	Benefits	Potential Concerns
Vinyl siding	• Durable • Maintenance free • Requires no additional site finishing • May be sourced locally • Potential reuse after expected life	• Health concerns related to manufacture and disposal of PVC
Wood siding	• Natural product, may be sourced locally • Low embodied energy • Durable when properly maintained • Raw material can come from sustainably managed forests	• Requires additional site finishing • Requires maintenance to ensure longevity
Stucco	• Natural material • May be sourced locally • Durable • Can be part of a thermal mass strategy	• Requires careful detailing to ensure longevity • Energy intensive manufacture
Fiber cement	• Durable when properly maintained • May contain recycled content • Potential reuse after expected life	• Requires additional site finishing • Health concerns of breathing dust during installation
Brick	• Extremely durable • May be produced locally • Requires no additional site finishing • Can be salvaged	• Energy intensive manufacture and distribution
Manufactured stone	• May contain recycled content • May be produced locally • Requires no additional site finishing	• Energy intensive manufacture • Requires careful detailing to ensure longevity
Plastic composite trim	• Requires no additional site finishing • Recyclable • Durable	• Health concerns with manufacture

ROOFING

In selecting roofing products, consider their longevity, energy use in manufacture, recyclable content, recyclability, and other factors (table 10.2).

TABLE 10.2 | Green properties of roofing alternatives

Roofing Product	Benefits	Potential Concerns
Asphalt shingles	• Durable (50-year warranty) • May be locally produced • Recyclable	• High embodied energy • Very few outlets for recycling
Certified Wood (shakes and shingles)	• Natural material • Relatively low energy for manufacture and transport • Reclaimed shakes available	
Clay tiles	• Durable • Can contain recycled content • May be produced locally • Can be 100% recyclable (Cradle-to-cradle–certified products available)	• Energy intensive manufacture and transport
Metal roofing	• Very durable • Recyclable	• Energy intensive during manufacture (recycled products less intensive)
Fiber cement shingles	• Very durable • May contain recycled content	• Manufacture presents environmental concerns • Precautions during installation necessary to protect workers from dust • Energy intensive manufacture and transport
Recycled roofing products (rubber; wood/vinyl)	• Recycled content (100% possible) • Durable (50-year warranty possible) • Lower embodied energy than virgin materials	

FLOORING

Flooring options may have various green characteristics:

- Natural, rapidly renewable, or recycled content
- Low energy use for manufacture and transport
- Low or no VOC emissions after product installation
- Low maintenance, including easy cleaning
- Durability
- Recyclability

When choosing flooring, consider water-based, no-VOC adhesives and finishes where applicable. Table 10.3 lists the green properties of a few flooring options.

TABLE 10.3 | Green properties of flooring alternatives

Flooring Material	Benefits	Potential Concerns
FSC-certified bamboo	• Durable • Natural, rapidly renewable material • Smooth surface, easy to clean	• Produced mostly in China • Prefinished floors are a better option for indoor air quality, but produce more emissions at the factory
Reclaimed flooring	• Durable • Natural material • Smooth surface, easy to clean • Post-consumer recycled content • Sourced locally	
FSC-certified wood flooring	• Durable • Natural material • Smooth surface, easy to clean • Can be sourced locally	• Often imported • Old-growth trees needed for production
Cork flooring	• Durable • Natural material • Smooth surface, easy to clean	• Softwood that can be dented • Most harvested in Europe
Linoleum (true linoleum)	• Durable • Natural, rapidly renewable raw materials (linseed oil, cork, jute, wood flour) • Smooth surface, easy to clean • Low VOCs • Pre- and postconsumer recycled content	
Recycled content and natural fiber carpeting	• CRI Green Label products have low VOCs • Recyclable • Type 6 nylon can be recycled back into carpet fiber repeatedly	• Natural fiber carpet may stain more readily than synthetics • Not often recycled, even though recycling is possible • May contribute to lower IAQ because of dust accumulation • Must consider environmental aspects of carpet pad
Recycled rubber	• Postconsumer and postindustrial recycled content	• Can off-gas and have odor long after installation • Rubber allergies
Concrete, tile, and stone	• Can be part of a passive solar design strategy • Contains natural materials • Durable • Resource efficient when slab is dual-purposed as finished floor • Smooth, easily cleanable surface • Can be sourced locally • Some can be salvaged at end of life	• Production and transport is energy intensive • Install with no-VOC adhesives, grout, and sealants to maintain IAQ

COUNTERTOPS

Countertop options for green homes continue to expand, as any trip through a builders' show will demonstrate. Natural materials and recycled content options are available. Products differ in their green properties, including energy to produce and transport to the jobsite, environmental emissions during production, and recycled content. Often, you will choose countertop materials based on aesthetics or customer preference. Table 10.4 shows some of the green properties of various countertop materials.

TABLE 10.4 | **Green properties of countertop options**

Material	Benefits	Potential Concerns
Concrete, tile, and stone*	• Contains natural materials • Durable • Smooth, easily cleanable surface • Can be sourced locally • Some can be salvaged at end of life	• Production and transport is energy intensive • Must be installed with no-VOC adhesives, grout, and sealants to maintain IAQ
Recycled content countertops (paper, stone)	• Postconsumer and postindustrial recycled content	• Limited distribution may require transport
Wood and high density cork	• Natural materials	• Not as durable as harder materials

*Although any natural rock has the potential to carry radon, and a 2008 study[i] showed radon levels from granite counters on par with background radiation levels, health hazards have not been proven. EPA's study, *Preliminary Granite Countertop Risk Assessment*[ii], has more information.

[i]Assessing Exposure to Radon and Radiation from Granite Countertops, prepared by Environmental Health and Engineering, Needham, Massachusetts. November 21, 2008.

[ii]Not available at the time of this writing.

CABINETRY

One longstanding concern with cabinetry in green building has been the urea-formaldehyde binders in the pressed wood. The Kitchen Cabinet Manufacturer's Association (KCMA) certifies cabinetry that uses wood products that meet California's stringent formaldehyde emissions limits. For information on KCMA's Environmental Stewardship Program certification, visit http: www.greencabinetsource.org. Other green cabinets use recycled or reclaimed wood, straw or other agricultural byproducts instead of wood, and low-VOC finishes. Or you can simply swap out shelving for a few cabinets to reduce material use.

INTERIOR AND EXTERIOR MILLWORK

For interior and exterior millwork, alternatives to old-growth lumber include FSC certified lumber, reclaimed lumber, wood/plastic composite lumber (typically made from wood industrial waste and recycled plastic), such as Trex, and finger-jointed/engineered trim that uses relatively young, fast-growing tree species. FSC's website (http://www.fscus.org) offers a list of certified building product retailers, many of which also carry finger-jointed exterior trim. Reclaimed lumber is found at local or online architectural salvage stores.

PAINT, COATINGS, AND SEALANTS

Architectural coatings and sealants help preserve a home's durability. To align with project goals, many green home builders select VOC-free (or low-VOC) water-based paints, adhesives, and caulk. Some or most VOCs will dissipate as the paint or sealant dries and your ventilation system will mitigate VOCs. However, when you consider the small incremental cost of using these alternatives compared with others, VOC-free or low-VOC alternatives are worth it not just to preserve the environment but also to enhance your company's image with home buyers.

VOCs are just one consideration in choosing paint. Paints that meet the *Green Seal Environmental Leadership Standard for Paint* (GS-11) offer other benefits. The GS-11 standard places stricter limits on VOCs for paint with colorant (EPA emission limits do not address pigments), bans several chemicals (like *toxic metals* and *tributyltins*) commonly used in paints, and establishes environmental criteria for packing and labeling.

For chemically sensitive individuals, natural paint with innocuous ingredients and trade names like Safecoat, milk paint, food paint, and clay paint are available. You can also purchase recycled-content latex paints but these can be difficult to find.

Although adhesives and caulks, like paints, fall under VOC compliance regulations, these rules vary greatly according to how you are using the products. The environmentally preferable alternatives are VOC compliant and solvent free. As with all caulks and adhesives, specify products that are rated for performance under the intended application. For example, APA-AFG 01 applies to subfloor adhesive, and ASTM

D3498 is the standard for strength and durability of bonding plywood to lumber framing. Also, when gluing OSB panels with sealed surfaces and edges (like Advantech), check the panel manufacturer's installation instructions. They may require polyurethane- or solvent-based glues.

HARDSCAPING

Like other finishes, the variety of products that are more sustainable than conventional materials continues to grow. These products can employ recycled materials, improve durability, reduce maintenance, use local and bio-based materials, and help rainwater percolate underground. Table 10.5 compares the green features for some hardscaping options.

TABLE 10.5 | Green properties of hardscaping materials

Material	Benefits	Potential Concerns
Certified wood decking	• Sourced from sustainably harvested lumber • Reusable/biodegradable • May be sourced locally • Relatively low embodied energy	• More natural resource intensive than recycled products
Reclaimed lumber wood decking	• Repurposed material	• Supply very limited
Wood/plastic composite decking	• Made from postconsumer and postindustrial recycled materials • Durable • Low-maintenance	
Pervious pavers	• Promotes natural rainwater infiltration	• Manufacture and transport is energy intensive
Permeable asphalt	• Promotes natural rainwater infiltration	• Manufacture and transport is energy intensive
Pervious concrete	• Promotes natural rainwater infiltration • May be sourced locally	• Maintain porosity by preventing dirt infiltration (through grading and occasional sweeping/vacuuming) • Energy intensive during manufacture and transport
Stone	• Natural product • May be locally sourced	
Recycled content retaining walls (CMUs, concrete)	• May be locally sourced • Durable	• Health concerns with manufacture

KEEPING UP WITH GREEN

The variety of green finishes is rapidly evolving. Keep up with the latest alternatives by attending building shows, browsing through magazines and green building supply stores, and talking with your customers. Green products don't necessarily cost more, but keeping up with new products requires an investment of time and professional education.

Subscribe to trade and consumer magazines and online resources like *EcoHome, Green Building Advisor, GreenBuilder,* and *Natural Home and Garden* to keep up with product developments between trade shows. Visiting green building supply stores and talking with other builders and customers can also help you stay current.

11

QUALITY CONTROL AND CONSTRUCTION MANAGEMENT

Even the most carefully designed high performance home isn't guaranteed, once built, to perform exceptionally. You must meticulously execute the plans for a home so it performs as it was designed to. This is where quality assurance (QA) enters the process. Quality control in the manufacturing process is well established. *ISO-9000*, *Six Sigma*, and other programs provide a roadmap to execute QA strategies. As every builder knows, building homes presents unique challenges to QA that distinguish construction from manufacturing on a factory floor:

- The environment during assembly is difficult or impossible to control. For example, construction crews encounter unfavorable weather conditions such as precipitation, wind, and extreme temperatures that affect both the process and building materials.

- Homes, even those built by "production builders," are not typically mass produced. To the contrary, many homes are one-off designs, unlike factory-made products that may be identical or at least comprise identical parts.

- Typically, many trade contractors representing different companies contribute to a home's construction. Controlling the final product, then, requires extra attention to specifications and processes like scheduling, communication, and follow-up.

Despite these challenges, many builders and other companies in the residential construction industry provide shining examples of how to apply quality principles to home building. The National Housing Quality Award has recognized some of these companies for their efforts to reduce defects and warranty costs, improve customer satisfaction, and improve a company's bottom line.[66]

WHAT IS QA?

QA is the systematic process of ensuring that a product, project, service, or organization meets pre-established standards. A builder's quality indicators might include the following:

- On-time delivery
- Profit or other economic indicators
- Costs (e.g., inspection, warranty, training)
- Defects at closing
- Number of warranty calls
- Satisfaction (e.g., customer, trade contractor, employee)
- Referral rates

Quality indicators for high-performing homes might include:

- Diagnostic test results including duct leakage, blower door test results, pressure differentials, temperature differentials, and combustion safety
- Home energy ratings (HERS Index)
- Green building ratings and certifications

When you incorporate QA mechanisms into a high-performing home, you can help ensure high levels of comfort, efficiency, and durability in addition to conventional measures of home quality such as plumb, level, and square.

National Housing Quality Award-winning companies have reported the following results after implementing quality initiatives:

- 98% of homes having zero defects at closing
- 12–20% increase in gross margins
- 15–50% decrease in *cycle time*
- 93–97% customer satisfaction rates
- 94% trade contractor satisfaction rates
- 95% employee satisfaction rates[67]

QUALITY MANAGEMENT TOOLS

Any business can adopt and customize a QA process to meet its needs. You can use scopes of work to set expectations for trade contractors and checklists to help ensure the jobsite is ready and each task is completed satisfactorily.

A four-page guide to best practices for planning, design, marketing, customer relations, and self-assessment derived from National Housing Quality Award winners is available at http://www.ibacos.com.

The Building America program specifies QA practices for high performance home building. The program offers sample scopes of work, inspection checklists, and educational information. A DOE Challenge Home Quality Criteria (BCQC) Support Document (http://tinyurl.com/BCsupportdoc) includes diagrams of proper installation techniques for high performance home features, sample job-ready checklists, inspection forms, and job complete forms to help ensure quality installations. The BCQC Support Document is an essential tool for all builders who want to meet quality standards for the energy efficiency, resource efficiency, IAQ, and durability of high performance homes.

In addition, online resources for detailed high performance scopes of work, job-ready checklists, and job-complete forms for several construction tasks, including excavation and backfill, basement foundations, slabs, rigid sheathing, and weather-resistant barriers are available.[68] Figure 11.1 shows a sample inspection list for slab foundations.

At the corporate level, the QA process begins by forming a quality team with an effective leader. The team oversees the process, assigns responsibilities, establishes quality standards, adopts quality management tools, measures effectiveness, and takes corrective action as needed.

The QA process in ISO 9000 consists of four steps: plan, do, check, and act.[69] Here's how to apply these steps in building a high performance home:

1. **Plan.** Set goals, determine your quality standards, establish performance metrics, formulate a budget that will be allocated for the QA process, and assign responsibilities. Make the plan comprehensive by involving as many stakeholders as possible.

FIGURE 11.1 | Sample inspection checklist

This inspection checklist for slab foundations, developed for the Building America program, is one of many quality tools available free online. *(NAHB Research Center)*

(Insert Builder/Trade Contractor Company Name & Logo)

Interior Building Slab Job Complete Checklist

Upon completion of work, the trade and builder must verify proper installation and sign off on all items as acceptable. Correction items needing correction must be re-inspected and confirmed-[BCQC-1]

Builder:	Lot:	Subdivision: Date of initial inspection
Builder Supervisor Name:		Crew Leader Name:

Instructions: Check box if OK, circle box if correction is needed. Note specific correction action required. When corrected, then check box.

		Trade Foreman OK	Trade Super OK	Builder OK/NA	Explain Needed Corrections List Date Corrected
	Describe Current Hotspot:	☐	☐	☐	
1.	Gravel sub base, specified insulation, and polyethylene are installed per specification and plan. [BCQC-23]	☐	☐	☐	
2.	Capillary break, insulation and polyethylene) were installed per plan between footing and wall. [BCQC-23]	☐	☐	☐	
3.	Drain tile, washed stone, filter fabric, sump crock is installed and connected per plan.	☐	☐	☐	
4.	Concrete delivery tickets show correct mix and are kept.	☐	☐	☐	
5.	Results of slump test(s) are noted.	☐	☐	☐	
6.	Slab level without high or low spots, pitting, or other surface defects.	☐	☐	☐	
7.	Slab elevation correct per site plan and dimensions correct per plans.	☐	☐	☐	
8.	Control joints installed per plan.	☐	☐	☐	
9.	Anchor bolts, tie-downs or other hardware are per plan.	☐	☐	☐	
10.	Forms stripped after concrete set up.	☐	☐	☐	
11.	No visible spalling of new concrete	☐	☐	☐	
12.	Site is cleaned per specifications.	☐	☐	☐	
13.	Concrete splatters and spills cleaned up.	☐	☐	☐	
14.	Streets are free of mud or concrete.	☐	☐	☐	
15.	All forms and excess materials are removed from the site.	☐	☐	☐	
16.	All debris in the proper recycling receptacle(s).	☐	☐	☐	
17.	Tolerances/specifications provided per scope.	☐	☐	☐	
	Additional Correction Items:				
			☐	☐	
			☐	☐	
Sign here to acknowledge that all above items have been completed	Trade Signature: Builder Signature:			Date: Date:	

Disclaimer

Neither the NAHB Research Center, Inc., nor any person acting on its behalf, makes any warranty, express or implied, with respect to the use of any information, apparatus, method, or process disclosed in this publication or that such use may not infringe privately owned rights, or assumes any liabilities with respect to the use of, or for damages resulting from the use of, any information, apparatus, method or process disclosed in this publication, or is responsible for statements made or opinions expressed by individual authors.

Stakeholders include not only upper management and office personnel, but also workers in the field on jobsites.

2. **Do.** Execute your plan on the jobsite and at the office. Standards and performance metrics help ensure quality installation and production efficiency. Office personnel will be responsible for measuring results. If you need to alter plans or scopes of work, this is the time to do it. Develop checklists and train staff and trade contractors. Training may be needed on changes to construction methods and materials, the use of checklists, and procedures for delivering QA information to the office.

3. **Check.** Use the inspection checklists developed in step 2 to document what is happening on the jobsite and determine whether workers are meeting jobsite performance objectives. The checklists will help you measure results and identify recurring problems. Identify who will be responsible for reviewing checklists. The "check" step is often neglected but, like all other steps, is key to a program's success.

4. **Act.** Make improvements as needed. Use *hot-spot* training to help trade contractors learn your quality standards and the details of how you want them to perform their work, to change materials if needed, or to change practices to resolve problems. If jobs are not meeting performance expectations, go back to the planning stage to develop solutions and revise plans for future projects. For example, if cycle time is too long, the team can examine whether a flaw in the plan is creating construction delays and can develop solutions to expedite future construction.

> The NGBS certification program and other green building certification programs use inspection checklists and networks of third parties to verify green construction. These checklists are an essential component of ensuring program compliance. You can use them in your internal QA process.

QA Roadmap for High Performance Residential Buildings offers 7 steps to QA:[70]

1. **Review past construction methods and assess performance.** Analyze current construction to reveal baseline results. Convene a multidisciplinary team to assess both the technical and business aspects of construction. The team conducts walk-throughs and performance tests of new homes and analyzes office data. Then, the team establishes baseline performance for cycle time, profit margin, number and frequency of defects, blower door test results, recurrent callback issues, warranty costs, and other indicators.

2. **Set performance goals.** With baseline indicators established, you can set performance goals for performance test results, cycle time, construction defects, warranty costs, profit margin, home buyer satisfaction, or other critical areas. Communicate these goals to staff and trade contractors.

3. **Determine, and adjust for, best practices to achieve performance goals.** Revise construction plans, specifications, contracts, and scopes of work as needed to accommodate new methods or materials. Use construction documents to precisely communicate new construction procedures and performance expectations. Reinforce those expectations verbally and in writing to staff members and trade contractors.

4. **Train field personnel and trade contractors.** Address new processes and products including storage, handling, and installation. Begin with training for managers, site supervisors, sales, and customer service staff. Next, expand the circle to include trade contractors in training with staff. This systematic approach to training will encourage finding solutions to potential problems before establishing new procedures on the jobsite. Finally, continue to support training of trades while they are learning new methods on the job.

5. **Evaluate success or failure by inspecting the jobsite.** Use verification resources such as the Building America Quality Control Checklist.[71] Verifying performance of a green home will add inspection criteria and testing to a jobsite supervisor's standard routine. For example, he or she may need to verify whole-house air tightness as part of this step.

6. **Commission the building. Systematically verify the operation of each system according to the quality standards selected for the QA process.** This step entails more than performing discrete tests like blower door, duct blaster, and room-by-room airflow. As discussed in chapter 9, commissioning comprises start-up procedures for testing each mechanical and electrical system within the home. ACCA Standard 9 HVAC Quality Installation Verification Protocols is a commissioning procedure for HVAC systems.[72] It contains sample commissioning reports and forms that you can adapt for your QA process. Test the proper operation of each of the home's systems in this step, including thermostats, lighting controls, water heating, ventilation, and space heating and cooling systems.

7. **Perform a post-construction performance evaluation.** Post-construction evaluation examines progress toward key

performance goals. Tasks for this step depend on performance goals established in step 2. For example, you can administer customer satisfaction surveys, perform long-term energy monitoring or utility bill analysis, and examine construction defects and rework.

PLAN

To begin the QA process, assemble a team of executives, field personnel, sales staff, budget managers, green certification program verifiers or consultants, and key trade contractors such as framers, electrical and mechanical contractors, and roofing and siding companies. The team will set performance standards and goals for the QA process. They will also establish a budget for the QA process, assign key responsibilities such as inspections, analyze data from the jobsite (checklists) and from the office (cycle time, *callbacks*, economic indicators) to determine whether goals are being met, to review past performance issues, and to plan and develop new training for employees and trade contractors.

The team can also develop a quality manual to be used company-wide that describes the company's QA system, explains standards, outlines scopes of work, and specifies tools that are part of the QA process. The quality manual is valuable for training employees on their roles and responsibilities in the QA process.

Establishing Quality Standards

A high performance home must meet quality standards for a wide range of products, systems, applications, and installations, from windows to drywall. Although building codes set minimum requirements, many professional and trade associations have established higher benchmarks for quality installation procedures and construction tolerances.

You can download a sample home building company quality manual (http://tinyurl.com/builderqualitymanual) from the NAHB Research Center (http://www.nahbrc.org). The NAHB Research Center's Toolbase site also lists documents to include in a quality manual (http://tinyurl.com/documentlist).

Instead of reinventing the wheel, you can adopt these existing standards or customize them as needed to align with your scopes of work and brand promise.

Quality standards for high performance homes extend over and above conventional home-building metrics like construction cycle time, budget, and defects. High performance quality standards also include metrics like blower door and duct leakage test results.

Resources

These resources discuss expectations for high performance homes:

- ENERGY STAR New Homes program
  http://www.energystar.gov
- NGBS
  http://www.nahbgreen.org
- Air Conditioning Contractors Association *HVAC Quality Installation Specification (ANSI /ACCA 5 QI)*
  https://www.acca.org/Files/?id=693
- *APA Engineered Wood Construction Guide* (American Plywood Association) http://tinyurl.com/APAregister
- *Recommendations for Installation in Residential and other Light Frame Construction* (http://www.nchh.org/Portals/0/Contents/Article0443.pdf) and *25 Checkpoints for Inspecting Insulation Jobs* (North American Insulation Manufacturers Association) http://www.naima.org/publications/BI497.PDF
- *Vinyl Siding Installation Manual* (Vinyl Siding Institute)
  http://www.vinylsiding.org/installation/manual/
- *Quality Control Guidelines for the Application of Asphalt Shingle Roof Systems* (joint publication of the Asphalt Roofing Manufacturers Association and National Roofing Contractors Association) http://tinyurl.com/qcasrs
- *Standard Practice and Installation Guides*[73] (American Architectural Manufacturers' Association)
- *ANSI A108/A118/A136.1 Specifications for the Installation of Ceramic Tile* (Tile Council of North America)
  http://tinyurl.com/tilespecs

- Builder Notes and Technical Notes from the Brick Industry Association (e.g., Tech Note 7B *Water Penetration Resistance*—Construction and Workmanship) http://tinyurl.com/bianotes

- Application and Finishing of Gypsum Panel Products (Gypsum Association) http://tinyurl.com/gypsumapps

- *Responsibilities for Inspection and Acceptance of Surfaces Prior to Painting and Decorating* (Painting and Decorating Contractors of America) http://tinyurl.com/paintsurfaces

Budgeting for QA

Consider the following needs in your QA budget:

- Labor expense for jobsite inspections, including third-party inspections, and documentation

- Time for planning and development (can vary greatly depending on the scope of your QA program)

- Development of training materials and time for training staff and trade contractors on new processes and products

- Materials such as inspection checklists, standards and specifications, new construction drawings

- How new methods or materials affect trade contractor construction schedules, budgets, and training requirements

- Review of inspection checklists and other documentation to identify areas that are working well and those that need improvement

Although the cost of implementing a quality program may seem daunting, the cost of not implementing the program is probably greater. For example, a 2006 study[74] of thousands of homes showed that builders spent, on average, $5,400 to correct defects on each home. Builders have reported cutting the number of defects in half by using QA practices.[75] Moreover, 80% of trade contractors who participated in the National Housing Quality certification program reported a reduction in callbacks. Many builders with successful QA programs attribute enviably high business results to their programs. For example, Veridian Homes of Madison, Wisconsin, a leader in the QA process for high performance homes, has a profit margin in the top 25% of the NAHB-sponsored Builder 20 groups. Grayson Homes of Maryland reported 98% of its homes were defect-free at closing and a 9% increase in net profits because of the company's QA process.[76]

DO

"Do" is the action step. The QA process is important across the entire organization and should include management, site supervisors, sales and marketing staff, accounting, and customer service. Activities may include

- company-wide training on the QA process to stress the importance of developing a corporate culture of quality and to assign key responsibilities;

- reviewing new QA tools such as inspection checklists, scopes of work, and processes, with people who will use them;

- organizing trade contractor meetings to introduce the QA process and to provide training on new materials and methods;

- following up on the company-wide training with on-site review; and

- using QA checklists to inspect work and provide immediate feedback to crew members.

BSC BUILDING AMERICA QC CHECKLIST

The BSC Building America Quality Control Checklist (http://www.buildingscience.com) is probably one of the simplest tools for keeping track of high performance home items while on the jobsite. Click on "information" and search for the title. The document includes an appendix that describes the checklist items and includes helpful diagrams.

If trades resist the QA process, you may need to seek other trade partners. You are in charge of setting expectations and your trade contractors are responsible for meeting them.

CHECK

Documentation is essential to a successful QA program. By documenting what is happening—through job-ready and inspection forms, third-party inspections, performance testing, and construction schedule and budget tracking—a company has the information to determine if performance metrics are being met. For high performance homes, documentation often involves a third-party green home verifier, an ENERGY STAR rater, or both.

Documentation is only valuable if you are using it to improve outcomes. Use the data you gather to determine whether your quality efforts are improving results and if they are not, what changes you need to make to meet performance standards.

ACT

The success of your quality program depends on what you do with the data you gather. The process of resolving issues, often called continuous improvement, is the crux of the QA process that drives positive changes, and often, improved profitability. Home builders using quality processes refer to inspections, feedback from on-site crews, and performance indicators like test results, to identify opportunities to improve. Following are examples of general goals a builder might set after reviewing this data:

- Reduce defects and repair costs for items that are not conforming to quality metrics

- Reduce the time to build a home or complete a specific task

- Improve the value of homes by enhancing their designs or changing building materials

- Increase efficiency of the building process by making improvements in the field, in the office, or both

Through its National Housing Quality program, the NAHB Research Center

QUALITY INDICATORS FOR HIGH PERFORMANCE HOMES

You can measure quality for a high performance home using these indicators and others:

- On-time delivery

- Whole-house air leakage test results

- Duct leakage test results

- HERS Index

- Green building program rating

- Room by room airflow

- Pressure differentials across doorways

- Profit

- Expenses (e.g., for callbacks, inspection, training)

- Customer, employee, or trade contractor satisfaction ratings

- Customer referral rates

- Productivity rates (home completions per employee or other indicators)

- Number of defects at walk-through or closing

- Number of warranty calls

pioneered the application of QA processes to the home building industry. Builders have applied the center's six-step process to eliminate hot spots in high performance homes:[77]

1. **Review data** from inspection forms and performance test reports to identify opportunities for improvement (aka, hot-spot issues).

2. **Prioritize issues** (based on cost, performance test results, or other criteria you choose).

3. **Identify what's causing problems** and create solutions.

4. **Update processes, installation methods or materials, and scopes of work** to support the solutions. Use hot-spot training to rapidly deploy new methods to field crews.

5. **Implement a solution on a small scale** and measure its impact before deciding whether to scale it up.

6. **Monitor results and adjust your practices as needed.** Once you resolve an issue, move on to the next priority hot spot.

One way to identify hot spots is to keep track of callback items. If you discover a hot spot, you can train all trade contractors with a role in fixing the problem. Hot-spot training materials

- focus on a single issue;

- include photographs showing correct and incorrect ways to perform a task or procedure;

- use symbols to identify the right and wrong ways (e.g., red ✗, green ✓);

- use arrows to identify the precise area of interest on the photos; and

- use graphics rather than text to depict the problem. If text is needed, provide English and Spanish if necessary.

The NAHB Research Center (http://www.nahbrc.org) offers a sample training tool, sign-in sheet, and guide to the hot-spot strategy, *Hotspot Implementation Guide and Tools*, (http://tinyurl.com/hotspottraining) on its ToolBase website.

A sign-in sheet for hot-spot training sessions helps track who attended the training and who is accountable for getting the job done right in the future.

CONTINUOUS IMPROVEMENT

The QA process is designed as a perpetual feedback loop so that after you resolve issues in the fourth and "final" step, the process starts again. Companies that participate in QA are continually redefining goals, setting standards, training staff and trades, and identifying problem areas in the field and the office. Companies keep practicing QA because it helps them achieve positive business results.

The National Housing Quality Certified Builder program (http://www.nahbrc.com/builder/quality/index.aspx) helps companies "build it right the first time, every time," instead of wasting resources dealing with problems after construction. Call 888-602-4663 for more information about how to join the program.

12

MARKETING AND CUSTOMER RELATIONS

Consumers are learning more about green products and, at the same time, becoming wary of green claims. Given the pervasiveness of *greenwashing*, consumers rightfully want tangible evidence of a product's benefits. Fortunately for builders, green building and energy efficiency programs quantify a home's energy and environmental performance. Certification programs distinguish green homes from other homes. However, earning a green building certification for a home is just a first step toward selling it. If green homes are new and scarce in your market, you will need to educate home buyers about what green means in the home-building realm. On the other hand, if green homes have become commonplace in your market, you will need to convince potential home buyers that your green home features are superior to your competitors'. Regardless of the local market, customers usually want to know what certification means and what the benefits and features of your green homes are.

Although today's home buyers are well informed, they are looking at your house for a variety of reasons, including its location, design, neighborhood, and schools. Environmental benefits are among these other factors that can help you clinch the sale; they may or may not be the most important factors in your buyer's mind. As with all home features, translate your green home features into benefits you can state concisely, convincingly, and truthfully. Following are some practical benefits of owning a green home that may resonate with buyers:

How to Sell ICC700 Green Certified Homes (http://www.nahbgreen.org/Content/pdf/HowToSellICC700GreenCertifiedHome.pdf) includes a table that describes the benefits of popular green features from the National Green Building Standard. You can adapt the text to fit your marketing materials.

- **Comfort.** Green homes can be more comfortable to live in because they have even temperature distribution throughout the house and good humidity control.

- **Low operating costs.** Energy efficiency and durability can save money on heating, cooling, electricity, water, and maintenance.

- **Positive environmental impact.** They can feel good that their decision to own a green home is saving natural resources today and will continue to do so in the future.

You can include specific details about key green features in your marketing materials but keep these explanations brief. Focus on what will capture your particular segment of home buyers. For example, water conservation might be paramount in Tucson, energy efficiency in Minneapolis, and low maintenance wherever you are building for mature empty nesters.

Your sales staff must understand the advantages of your homes and be able to relate them so buyers comprehend them. In other words, emphasize the benefits and avoid building science jargon that might bore the average home buyer.

TIPS FOR GREEN MARKETING MATERIALS

- Emphasize the benefits of living in your homes.

- Be specific, but brief, in discussing green features.

- Know your audience.

- Avoid jargon.

- Train sales staff members to explain the benefits in simple terms and the features, if necessary, in technical terms.

- Validate marketing claims using green and energy efficiency ratings and certifications.

- Leverage your marketing budget by using materials developed by green building programs. For example, the NAHB Research Center offers consumer brochures about what Green Certified means (http://www.nahbgreen.org/Content/pdf/GreenConsumerGeneral.pdf).

You won't need to reinvent the wheel if you use turnkey marketing materials from green building programs. These free resources include flyers, yard signs, and brochures. They are professionally designed and written to explain what green is and what certification means.

SELLING ENERGY EFFICIENCY

Because it has tangible financial benefits, energy efficiency is the most obvious—and can be the easiest—route to a green home sale. However, you must prove your green homes are more efficient than your competitors' homes, because many buyers will assume that any new home is energy efficient, whether or not the home is green. You can reference your home's ratings from third-party performance testing, but be prepared to explain what the ratings mean: the average home shopper might not be familiar with a HERS Index. The ENERGY STAR logo is recognizable to most people, and programs that offer energy-use guarantees, such as Environments for Living, can help you sell energy efficiency as well.

According to J.D. Power and Associates, the number of new home owners who perceived their homes as environmentally friendly jumped from 31% in 2009 to 65% in 2010. More than half of the builders that the J.D. Power surveys encompassed represented to customers that their homes were green.[78] The actual number of certified green homes compared with all home starts suggests that many more builders are calling their homes green than are seeking green building certification. You can distinguish your homes, educate your market, and bolster your credibility by having an independent green home verifier certify your homes.

AWARDS PROGRAMS

If you are building energy-efficient green homes, consider entering one or all of the following awards programs:

- National Green Building Awards
- LEED for Homes Project of the Year
- ENERGY STAR Partner of the Year (New Homes)
- Local competitions

Awards programs are a great public relations tool to raise awareness and company prestige. They usually offer free public relations, such as online media or magazine exposure; plaques you can hang in your sales office or model home; and the right to call your company or home(s) "award winning" and attach an awards program logo to them.

OPEN HOUSES

Open houses offer a perfect opportunity to present and demonstrate the green features of your new homes. Point out these features with placards placed throughout the home. Always include takeaways that discuss the benefits of these features and the significance of green home certification. These brochures and other collateral will help buyers remember and distinguish your home from others they will tour in their search for a new home.

You can also participate in a local Parade of Homes to attract potential buyers. Many Parade guides now include map keys that identify green homes. A Parade tour provides a fantastic opportunity to make a lasting impression and to educate the community at large about your green homes.

THE NATIONAL GREEN BUILDING AWARDS

The National Green Building Awards recognize companies for excellence in residential green building design and construction. The prestigious awards honor home builders, remodelers, and others for advancing green building through innovative design and construction techniques, excellent educational programs, and successful advocacy efforts. Submissions do not need to be certified green, but must be scored to the National Green Building Standard. The awards, which are open to NAHB members and nonmembers, are announced at a ceremony at the National Green Building Conference and Expo. For more information, visit www.nahb.org/greenbuildingawards.

FIGURE 12.1 | 2010 National Green Building Award Winner

This Emerald Certified home was designed to cope with high ground water and for energy efficiency in the harsh winter climate of northern Michigan. It includes a southern orientation, wind breaks, a raised foundation with drain stone, and advanced framing. *(JR Construction Building & Design LLC)*

PROFESSIONAL WEBSITES

Your company website is the most important communication tool you have to showcase your homes. An effective website will compel visitors to visit your homes and communities or at least contact your sales staff.[79] It's not difficult to create a website that presents your company favorably to prospective home buyers:

- Consider hiring a professional designer to ensure your website is attractive, contains the content that will attract buyers and search engines like Google, is easy to use, and helps convert Internet traffic into home visitors.

- Describe the benefits of your green homes prominently in *plain language* on the *landing page*.

- Prominently display your professional designations, such as NAHB's Certified Green Professional (CGP), and awards and certifications that you have earned to lend credibility to your company.

- Link to technical information about your homes' green features for home buyers who want more details.

- Include professional photography but use *low-resolution* photos that will load quickly.

- Include a call to action like "contact us" and an easy-to-complete form so potential buyers can get more information.

ATTRACTING PRESS COVERAGE

Your company may be doing great things, but if nobody but you knows about them they will not sell homes or promote green home construction. One way to spread the word about your company, homes, and green construction methods is to court local media. Reporters and editors, especially those in smaller media markets, are constantly seeking news stories and feature stories that will engage readers, listeners, and viewers with something different. You can reach out with a brief (½ to 1 page) press release or an e-mail message. Make sure to follow up with a phone call so you can answer questions and explain why your idea is newsworthy. Depending on what else is occurring in your local market, your company's earning certification for its Nth green home might be enough to attract press coverage. Compared with paid advertising, developing media contacts is inexpensive, although it requires a time commitment. Because you are not paying for this coverage, consumers also may view press reports as more credible than paid advertising.

CUSTOMER RELATIONS

The first step to getting a referral is building good relationships with existing customers. These relationships often begin well before a sale and continue long afterward. For a green home builder, customer education is likely to be a big component of relationship building.

Flynner Building Company of Boise, Idaho, attracted press coverage for donating time and expertise to participating in the St. Jude's Dream Home Giveaway. The company received media buzz—and about 10,000 visitors toured the home—during a four-month comprehensive media campaign sponsored by the charity. For the project, Flynner Building Company elected to build a net-zero energy home, which raised $700,000 for the charity and generated valuable regional interest in the builder.

Builders who work with customers during the construction process have an advantage compared with those who sell inventory homes. In fact, when you are building a home for an existing customer, you will spend, on average, about 30 hours with him or her during planning and construction.[80] After a home is completed and sold, you can continue to cultivate good customer relations. Fortunately, builders have many opportunities to present information to home owners, whether in person during walk-throughs, during follow-up phone calls, in a printed home owner manual, or in marketing materials.

WALK-THROUGH(S)

Include at least two walk-throughs—pre-drywall and after closing—to help consumers understand how their new green home works and how to operate it. This will help them get the most from its high performance features.

Pre-Drywall Walk-Through

Show home buyers what is behind the walls of your green homes to help them understand their energy-saving features. Point out how insulation fills wall cavities, what the window efficiency ratings on the labels mean, air-sealing measures, advanced framing techniques, duct sealing and layout, duct airflow dampers, the benefits of engineered lumber, cutting-edge roof framing methods, house wrap installation and flashing techniques, plumbing layout, and other features that contribute to high performance.

High-volume builder Meritage Homes maintains a "deconstructed" model home in each of its neighborhoods. Outfitted with educational signs about duct systems, insulation, and other energy-saving features, these hands-on exhibits show potential buyers the principles of energy efficiency and green features in practice (fig. 12.2).

Refer to the inspection checklists you develop for your QA program to ensure you conduct a comprehensive tour. This is your opportunity to discuss the science behind your homes—seize it! Feel free to offer technical explanations to customers who want to learn more but practice talking about your green practices in plain language.

If you are building inventory homes, consider adding a pre-drywall walk-through on a yet-to-be-completed home for buyers who are purchasing prebuilt homes, even if they have already closed on their home. If a pre-drywall home is not available to view, you can still explain what's behind the walls using cutaway illustrations or mock-ups. You want home buyers to thoroughly understand what they're getting so they can communicate the benefits of your high performance homes to their networks of family, friends, coworkers, and others who might consider purchasing one of your homes.

FIGURE 12.2 | Framing tour with educational signs

A framing tour provides an ideal opportunity to showcase a home's performance benefits. *(Meritage Homes)*

Show off your product using displays, cutaways, mock-ups, and educational signs in offices and model homes. Colorado Dream Homes of Pagosa Springs, Colorado, has a mock-up of its unique double-stud wall system on permanent display in its sales office to visually promote its highly insulated wall construction technique (fig. 12.3).

FIGURE 12.3 | Model to demonstrate energy efficiency

Three-dimensional displays demonstrate energy efficiency concepts and how your product differs from your competitors'. *(Colorado Dream Homes)*

Pre-settlement Walk-Through

Use the pre-settlement walk-through to orient home owners to the systems in their home and how to operate them. This orientation will last longer than what you may be accustomed to. Because green home systems are more complex, orientation is crucial. You may need to conduct a typical walk-through and then schedule a second orientation to demonstrate how to operate the new green home. Consider following up with a postcard to remind home owners about required maintenance items like changing furnace filters and checking the radon vacuum gauge or call them to informally inquire about the home's performance. These simple customer relationship management (CRM) activities can help keep a home performing optimally and reassure your home buyers that you believe in your product's high quality. CRM software can help you automate follow-up, and books such as *Customer Service for Home Builders*[81] provide ideas and strategies.

High performance home features such as radon mitigation systems, high-efficiency air filters, and sophisticated lighting and HVAC controls probably require more explanation than conventional home features of the past. As you discuss how these green home features work during the pre-settlement walk-through, demonstrate how to operate them. Use a checklist to ensure you don't omit important items and provide the list as part of your home owner manual. Here are a few suggestions for handling the pre-settlement walk-through:

- Discuss the grading, landscaping, and gutter system that contribute to foundation moisture control.

- Operate high performance mechanical systems. Turn on faucets to demonstrate how quickly hot water arrives, operate dual-flush toilets, turn on the cooling and heating system, and demonstrate lighting controls and lighting design (e.g., task, ambient) features.

- Ask customers to notice how fresh the air smells inside the home (no "new house smell").

- Showcase the resource-efficient and beautiful finishes—the "wow" factor.

THE HOME OWNER MANUAL

Buyers will be overwhelmed during the closing process and after move in so they probably won't remember everything you've shared at the walk-through. Provide a comprehensive home owner manual they can refer to later. NGBS mandates that to have a home Green Certified, builders must provide a home owner manual. The program offers additional points toward certification for including green lifestyle information such as local public transit options and for explaining green lawn care and cleaning methods.

You don't need to create a home owner manual from scratch. *Your New Green Home and How to Take Care of It*[82] is a customizable template on CD for homes certified to the NGBS. Just indicate which features are included in the home and the template produces a manual for you.

POST-CLOSING CUSTOMER RELATIONSHIP MANAGEMENT

Many customers, of course, will not open their manual unless or until they believe there is a problem with their new home. You can avoid unnecessary warranty claims by sending maintenance reminders to your home owners. Customer relationship management (CRM) also includes the following steps:

- **Follow up with the home owner by phone.** Telephone the home owner after they have lived in the home for two weeks. This is an opportunity to maintain personal contact and proactively address questions and concerns.

- **Send a thank you note.** Thank you notes don't have to be elaborate to demonstrate that you appreciate the customer's purchase. This home is one of the biggest purchases of their lives. Let them know you are grateful for their business.

- **Send a welcome letter.** A welcome letter, separate from a thank you note, reminds home owners about scheduled maintenance and other information they can reference in their home owner manual.

- **Contact customers at scheduled intervals.** You can use CRM software to schedule reminders to follow up with letters or other personal contact, such as at 3 weeks, 30 days, 3 months, and 1 year.

- **Send newsletters.** A CRM system can generate newsletters to keep home owners up to date about your company's activities and achievements.

Although using a computer to manage customer relationships may seem counterintuitive, software can be helpful in automating customer relationship tasks. Builder-specific software can help you keep track of the construction process as well as customer touch points so you don't forget important follow-up activities.

If you are selling an inventory home, follow-up activities can help develop relationships. If you are building post contract, you can incorporate follow-up activities to build on the established relationship with the home owner and contribute to a positive experience that will help drive referrals from family, friends, and colleagues.

KEEP CUSTOMERS FOR LIFE

Getting customers is your primary goal, but keeping them (and keeping them happy) is a long-term investment in your company. It takes work to establish your credibility as a builder of quality green homes. Although marketing is essential for drawing potential buyers to your homes, and applying the building science discussed in this book will help you construct the best high performance home possible, solid CRM will help your company maintain its brand promise. Good relationships help generate referrals and repeat buyers.

GLOSSARY

ACH50. Air changes per hour at a pressure difference between indoors and outside of 50 Pascals. In home energy testing, this pressure difference is induced with a blower door and the ACH50 indicates the leakiness of a house. Leakiness can be expressed in other terms, but ACH50 is the most common and is referenced in the 2012 IRC.

alternating current (AC). Electrical charge which reverses direction periodically—60 times per second (or Hertz for U.S. residences). It is the type of power that is delivered to homes and businesses.

anemometer. An instrument that measures wind speed or airflow

array (photovoltaic). A collection of electrically connected photovoltaic panels

backdrafting. The potentially dangerous condition in which the by-products of combustion are drawn into a house because there is insufficient airflow for natural-draft combustion appliances.

bituminous. An asphalt- or bitumen-based waterproofing product

blower door. A device for measuring whole house air leakage

building envelope. The structural elements (walls, roof, floor, foundation) of a building that enclose conditioned space; the building shell.

building-integrated photovoltaic (BIPV) panels. Electricity-producing solar panels that also serve as a building element (e.g., roofing material)

building science. The study and body of scientific information related to the physics of a building including moisture transport and energy flows.

callback. A customer complaint, within the warranty period, for repair to be done at the builder's expense.

capillary suction. The ability of a liquid to flow against gravity in a small-diameter tube through the combination of surface tension within the liquid and adhesion between the liquid and tube

chimney effect. The condition wherein less dense warm air rises, causing higher pressure at the top of a house, and replacement air is naturally drafted inward from the lower portion of the house; can be a passive ventilation strategy.

Class I vapor retarder. Sheet polyethylene or non-perforated aluminum foil

Class II vapor retarder. kraft-faced fiberglass batts

Class III vapor retarder. latex or enamel paint[83]

color rendering index (CRI). A measure of a lamp's ability to accurately reproduce colors in comparison to their color under natural light. Expressed as a number between 0 and 100, with 100 being the most accurate.

color temperature (CT). A characteristic of visible light, expressed in the Kelvin scale, that refers to its hue. CT over 5,000K refers to bluish, cool colors whereas CT between 2,700K and 2,800 indicates warm, yellowish hues similar to incandescent light.

compost sock. A mesh tube filled with compost that is used to prevent sediment runoff

cycle time. The time required to deliver a home to a customer from the start of construction. May or may not include site development.

daylighting. Design elements that capitalize on sunlight to provide natural lighting within the home and minimize the use of artificial lighting

de-superheater. A device used in conjunction with a ground-source heat pump that collects excess thermal energy from superheated heat exchange fluid (also called refrigerant) to produce hot water for domestic use. Because it cools the superheated fluid, a de-superheater also enhances cooling efficiency.

dimensionally stable. The property of a building material in which its size remains relatively constant under varying environmental conditions (e.g., temperature and moisture)

direct current (DC). Electric current that flows in one direction. DC is produced by batteries and solar cells and is also used to charge batteries.

drip edge. Metal flashing applied to the edges of a roof to prevent water from getting between the underlayment and the roof sheathing and to help divert water from the roof system

dual-flush toilet. A toilet with the capability to flush using two different volumes. Typical volumes are 0.8 gallons for liquid waste and 1.6 gallons for solid waste.

elastomeric sprays. Spray-applied sealants that remain flexible after drying and form a barrier to air penetration.

embodied energy. All of the energy used to produce a material, including raw material extraction, manufacture, and transport.

energy recovery ventilator (ERV). Equipment that draws fresh outdoor air into a home and removes stale air from within while it exchanges latent (moisture) and sensible (temperature) energy between the two streams

evacuated tube. A type of solar thermal collector that encases the heat-absorbing surface in a vacuum within a glass tube. The vacuum improves thermal efficiency under cold ambient conditions.

expanded polystyrene (EPS). Rigid foam insulation manufactured from expandable polystyrene resin. EPS typically is white and has a beaded appearance. It is the material used in foam coolers and insulated cups.

extruded polystyrene (XPS). Rigid foam insulation manufactured using the extrusion process. Most commercially-available XPS is blue or pink.

flash seal. Using a thin layer of spray foam insulation to air seal a building while minimizing the cost of the insulating material

flat-plate. A type of solar thermal collector with a heat-absorbing surface that appears flat. Large heat absorbing fins extend from small tubes through which fluid is pumped to deliver the solar energy to a storage device.

flex duct. Flexible ductwork that is typically insulated on the exterior

flow hood (also called air capture hood). An electronic device for reading low-volume airflow through HVAC registers that is useful for balancing airflow and verifying that design airflows are realized.

frost-protected shallow foundations (FPSFs). A method for protecting foundations from frost heave without placing footers below the frost line

geoexchange. A ground-source heat pump

gpm. gallons per minute

gray water. Water that has been used within the home but has not been exposed to human waste, such as water that drains from the clothes washer and dishwasher.

greenwashing. Exaggerating the environmental benefits of a product or process

header hangers. Metal fasteners that can be used in lieu of jack studs to support a header

heating penalty. The additional energy required to condition incoming ventilation air

heat recovery ventilator (HRV). See energy recovery ventilator (ERV).

high-density polyethylene (HDPE). Rigid plastic that is highly durable and resists decomposition

high performance home. A home designed and built to be more energy-, resource-, and water-efficient, more durable, and to have fewer pollutant emissions than a home built to meet minimum code requirements.

hot spot. A recurring problem within the construction process

hydric soils. Soils that formed while saturated, flooded, or being ponded over a sufficient time period to starve the upper layer of oxygen during the growing season. Hydric soils indicate wetlands or other moisture issues that could negatively impact the durability of a structure.

hydronic. A system of heating or cooling a home by circulating fluid through a distribution system. Examples include baseboard heating and in-floor radiant heat.

hydrostatic force (hydrostatic pressure). The force of unmoving water

hygrothermal behavior. Behavior related to temperature and humidity

IC-rated. Recessed lighting fixture that is rated for contact with insulation

insulating concrete forms (ICFs). Stackable foam forms filled with reinforcing steel and concrete that provide both interior and exterior insulation

integrated-collector storage (ICS) systems. A solar thermal water heating system that integrates collection and storage. Used in climates where the potential for freezing is very low.

ISO-9000. A series of international standards for quality management systems

jack studs. The studs that support a header from below, transfer the load from the header to the floor system, and form the rough opening of a window or door.

jump ducts. Ducts that provide an air pathway across closed doorways by "jumping" over the doorway above the ceiling. When combined with central return pathways, jump ducts and transfer grilles balance pressure across doorways, eliminating the need for undercutting doors or providing returns from each bedroom.

kilowatt (kW). An instantaneous measure of energy consumption. Typically used to describe the electricity demand at a point in time. A kilowatt-hour (kWh) is equivalent to 1 kW over 1 hour. Forty 25-watt lightbulbs burning for one hour would consume 1 kWh of electricity.

ladder blocking. An OVE framing method in which interior partition walls are connected to exterior walls by horizontal ladder-rung-type framing between vertical studs on the exterior wall. This framing method provides a nailing surface for the connection while allowing space for continuous insulation in the exterior wall behind the ladder framing. It can reduce lumber usage and employ scrap lumber.

landing page. The page a visitor encounters first when he or she visits a website

latent heat. Energy that is released or absorbed by a substance during a phase change (e.g., condensation of water vapor to liquid water) at a constant temperature.

layering. A design strategy for functionally and aesthetically lighting a home. Layers include task lighting such as under-cabinet and desk lighting, ambient lighting, and accent lighting for decoration.

life cycle assessment (LCA). A method for calculating the environmental impact of a product or system based on its impact on the environment from manufacture to disposal.

low-resolution. A description for an image that indicates the picture quality is fine for a web page but that the image would be distorted if enlarged or used in printed materials

lumens per watt (LPW). The ratio of the light output of a lamp (in lumens) to the amount of electricity input. The higher the LPW, the more efficient the bulb is.

methylene diphenyl diisocyanate (MDI). A chemical used in the manufacture of polyurethane

miscellaneous electric loads. Electric uses in a home for other than major appliances, space conditioning, and lighting. These loads typically include small appliances and consumer electronics.

OG-300 certified systems. Turnkey solar water heating systems that meet requirements of the Solar Rating and Certification Corporation's solar water heating system rating and certification program

Optimum Value Engineering (OVE). A set of methods for building structurally sound homes using less framing lumber than is typical

passive infrared. A technology that detects motion by using a sensor that measures thermal energy. It is used to control lighting or other home features.

passive solar. A building design that uses structural elements of a building to heat and cool it without mechanical equipment

phenol-formaldehyde. A resin used in exterior adhesives for its water-resistant properties

photocells. Lighting controls that detect light levels to activate lighting

photovoltaic (PV). Semiconductor materials that convert a portion of the incoming solar radiation into DC electricity

plain language. Language that the general public can comprehend and which does not include construction jargon

plug loads. See miscellaneous electric loads

polyethylene. Plastic film, often referred to by the brand name Visqueen. It is impermeable to water vapor and is rated by its thickness in mils.

polyisocyanurate (polyiso). A closed-cell insulating foam that can be spray applied. Also, a rigid foam panel. Panels typically can have a variety of facings, such as reflective foil or structural board.

psi. Pounds per square inch. A measurement of force.

radiant barrier. A material that has a low emissivity and, therefore, does not permit most thermal energy from being radiated outward from its surface.

radiant heat. Heat transmitted directly to objects without necessarily heating the surrounding air. An example of radiant energy is the warmth you feel from a fireplace that you don't feel if someone stands between you and the fire.

rain garden. A planted depression that allows rainwater runoff from impervious urban areas like roofs, driveways, walkways, parking lots, and compacted lawn areas to be absorbed, reducing the amount of polluted water rushing into creeks and streams.

rain screen. A method for attaching siding to maintain adequate space between the siding and a moisture-resistant sheathing or house wrap. This allows for drainage and ventilation, and reduces the pressure difference across the siding that drives moisture toward the sheathing.

rainwater catchment system (rainwater harvesting system). A system for collecting and storing rainwater for home use indoors or outdoors

raised-heel truss. Pitched roof truss in which the top angled chord and bottom horizontal chord do not meet at a point over the exterior wall. Instead, they are joined by a vertical member that adds vertical height over the exterior wall to allow full-height insulation to extend completely over exterior walls in this commonly underinsulated area.

R-value. Resistance to heat transfer. The higher the R-value, the better the material is at preventing heat transfer.

scope of work. A description of the work a trade contractor is supposed to accomplish

semi-vapor-permeable. Having a permeance greater than 0.1

sensible heat. Thermal energy that results in a change in temperature

sill sealer. A material between the top of a foundation wall and the exterior wall that forms an air seal and prevents capillary movement of water between the foundation and the wall system

silt fences. Mesh fences placed around the exterior of a building site to prevent storm water runoff from the site

Six Sigma. A quality management process, originally developed by Motorola, for ensuring that high standards of quality can be met. Six Sigma refers to the statistical notation for a process in which 99.99966% of the products are expected to be free of defects.

soil stability study. A test which evaluates soil resistance to erosion or landslides and which measures the stability of the soil when exposed to rapid wetting. Stability is affected by particle size and the chemical constituents of the soil affect stability.

solar energy factor (SEF). An SRCC rating of performance for solar water heating systems, defined as the solar energy delivered to the hot water divided by the sum of the purchased energy input to the entire system (e.g., electricity for pumps and auxiliary hot water energy use) at test conditions.

solar fraction (SF). The fraction of energy delivered that is expected to be provided by solar, rather than auxiliary, energy.

solar heat gain coefficient. The ability of a window to allow sunlight to transmit heat to the interior. SHGC is a number between 0 and 1 representing the fraction of incident solar radiation that is transmitted as heat to the interior. The lower the SHGC, the less solar heat gain that will occur.

solar reflectance. The ability of an object to reflect the incident solar energy. Energy not reflected is absorbed.

source energy. The equivalent amount of energy needed from a fuel source to obtain energy at the site (site energy). Source energy accounts for the inefficiency and losses in generating and distributing energy. For electricity, it takes more than three units of source energy for every unit of electricity delivered to a home.

Spray Polyurethane Foam Association (SPFA). An industry organization dedicated to promoting the use of spray polyurethane foams

standby mode. The state of a remotely controlled electronic device when it has been powered off by a remote, but not at the electrical circuit. All electronic devices draw some energy in standby mode.

steep slope. Typically a slope of 15%–25%, but the definition varies by jurisdiction.

structural insulated panels (SIPs). Structural wall and roof panels that have an insulating foam core (typically EPS) between two structural facings (typically oriented strand board, or OSB).

sun-tempered design. A design strategy that uses the sun's energy to partially light and heat a home. It must be supplemented by energy from other sources.

sweating. Condensation on the outside surface of pipes. It occurs when ambient air is at a sufficiently high temperature and relative humidity to condense on the relatively cool surface of the pipe.

tankless water heater. Equipment that heats water instantaneously and does not contain a reservoir of preheated water

thermal bridges. Areas in a building envelope through which energy can transfer more readily between interior and exterior surfaces than it does across most of the building's exterior.

thermal emittance. The ratio of the thermal energy (in the infrared) emitted from a material to the emittance of a perfect thermal emitter. Thermal emittance is expressed as a number between 0 and 1, with 0 being a perfect reflector of thermal energy and 1 being a perfect absorber (and emitter) of thermal energy.

thermosiphoning. The natural, convective movement of a fluid that occurs because of differences in temperature. A warmer fluid is less dense than a cooler fluid and, hence, rises above a cooler fluid. Thermosiphoning can cause unnecessary energy losses in hot water systems.

toxic metals. Metals known to cause illness in, or be carcinogenic to, humans

transfer grilles. Devices that allow air to flow freely across interior doorways to provide a return air path to central return ducts. Transfer grilles can be a key component of a low-cost, high performance air distribution system.

tributyltins. A group of organic pollutants considered toxic and to have negative effects on humans and the environment

U-factor. A measurement of a window's ability to conduct thermal energy. The lower the U-factor, the better the window is at insulation. A U-factor of 1 indicates the window is a perfect conductor of thermal energy and a U-factor of 0 indicates a perfect insulator.

ultrasonic. Outside the range of human hearing

UV radiation. Radiation from the sun in the range of 4–400 nanometers. It causes sunburn but also helps humans generate vitamin D. When it penetrates windows, it can fade home finishes and furnishings.

vegetated swales. Depressions designed into the landscaping to prevent storm water runoff and create a path that maximizes the time storm water is in the swale so fewer pollutants enter local waterways.

volatile organic compound (VOC). A harmful chemical containing carbon that readily evaporates (or sublimates) into surrounding air at room temperature

wind washing. The movement of relatively cold exterior air through fibrous insulation material that causes convective heat loss through the material and reduces its overall insulating value

xeriscaping. Landscaping designed to resist drought (e.g., to not rely on watering)

NOTES

[1] Organizations that, in my opinion, excel at getting valuable building science information out to the public on the World Wide Web include Building Science Corporation, Green Building Advisor, NAHB Research Center, manufacturer's organizations, Advanced Energy, Southface Institute, Consol, Energy and Environmental Building Association, Affordable Comfort, U.S. Green Building Council, USDA Forest Products Laboratory, the Cooperative Extension Service, Building America, ENERGY STAR, high school and college residential construction programs.

[2] LEED for Homes (http://www.usgbc.org/homes) is another popular national certification program for green homes. Because NGBS offers more flexibility for builders (particularly those getting started with green building), was adopted as a national standard through the ANSI consensus process, and has been found to be less expensive to achieve certification, this book focuses on NGBS as a more practical alternative for builders of conventional homes.

[3] U.S. Energy Information Administration, *Annual Energy Review 2009*. Table 2.4, DOE/EIA-0384m August 2010, http://www.eia.gov/aer.

[4] *Costs and Benefits of Green Affordable Housing*, 2005, report from New Ecology Inc. and the Tellus Institute. Available for purchase from http://newecology.org/costs-and-benefits-green-affordable-housing-study.

[5] *Green Home Building Rating Systems – A Sample Comparison*, Upper Marlboro, MD: NAHB Research Center, March 2008, http://www.nahbgreen.org/Content/pdf/GreenHomeRatingComparison.pdf.

[6] Numerous sources (e.g., Utah Governor's Office of Planning and Budget, Steep Slopes, http://planning.utah.gov/criticallands/Critical%20Lands%20PDFs/steepslopes.pdf; State of New Jersey, Steep Slope Model Ordinance, 2008, http://www.state.nj.us/dep/watershedmgt/DOCS/WQMP/steep_slope_model_ordinance062408.pdf).

[7] *Revisions to Quality Management Products: Four Scopes of Work for High Performance Homes*, Upper Marlboro, MD: NAHB Research Center, 2008. http://www.toolbase.org/PDF/BestPractices/ScopesofWork.pdf.

[8] See note 7.

[9] National Ready Mixed Concrete Association fact sheet CIP-11, Curing In-Place Concrete, http://www.nrmca.org/aboutconcrete/cips/11p.pdf.

[10] International Code Council, *2012 International Residential Code For One-and-Two Family Dwellings*, Clifton Park, NY: Delmar Cengage Learning, 2011.

[11] Building Science Corporation, Basement Insulation, Information Sheet 511, 2009. *http://www.buildingscience.com/documents/information-sheets/5-thermal-control/basement-insulation*.

[12] International Code Council, *2012 International Energy Conservation Code*, Clifton Park, NY: Delmar Cengage Learning, July 2011.

[13] PS1 plywood and PS2 OSB panels easily meet Japanese Agricultural Standards and Europe's formaldehyde emissions limits.

[14] *Thermal Design and Code Compliance for Cold-Formed Steel Walls*, Washington, DC: Steel Framing Alliance, August 2008.

[15] Oak Ridge National Laboratory, WUFI software, available from http://www.ornl.gov/sci/ees/etsd/btric/wufi_software.shtml.

[16] Bludau, C., D. Zirkelbach, and H.M. Künzel, 2008, *Condensation Problems in Cool Roofs*, Proceedings of the 11th International Conference on Durability of Building Materials and Components, Istanbul, Turkey. May 11–14, 2008.

[17] Oak Ridge National Laboratory (ORNL), *Radiant Barrier Fact Sheet*, http://www.ornl.gov/sci/ees/etsd/btric/RadiantBarrier/. Accessed February 23, 2011.

[18] See note 17.

[19] Straube, J., J. Smegal and J. Smith, 2010, Moisture-Safe Unvented Wood Roof Systems, Research Report 1001, http://www.buildingscience.com/documents/reports/rr-1001-moisture-safe-unvented-wood-roof-systems.

[20] Hendron, R., S. Farrar-Nagy, R. Anderson, P. Reeves, and E. Hancock, "Thermal Performance of Unvented Attics in Hot-Dry Climates: Results from Building America," Golden, Colorado: National Renewable Energy Laboratory, 2003, http://www.nrel.gov/docs/fy03osti/32827.pdf.

[21] William Turley, executive director of Construction Materials Recycling Association, August 2011, personal communication.

[22] For example, fiberglass SeriousWindows can have a U-factor as low as 0.09. Double hung windows are available as low as U-0.14.

[23] For a review of climate-specific design of vapor barriers in wall systems, see Building Science Corporation report, "Vapor Barriers and Wall Design," http://www.buildingscience.com/documents/reports/rr-0410-vapor-barriers-and-wall-design.

[24] Building Science Corporation, Insulations, Sheathing, and Vapor Retarders Research Report 0412 (2004), http://www.buildingscience.com/documents/reports/rr-0412-insulations-sheathings-and-vapor-retarders.

[25] Chapter 7 of the 2012 IRC outlines vapor barrier requirements and does not permit vapor barriers in wall construction in climate zones 1, 2, and 3.

[26] As defined in Appendix A of RESNET's 2006 Mortgage Industry National Home Energy Rating Systems Standards, Appendix A, http://www.resnet.us/standards/RESNET_Mortgage_Industry_National_HERS_Standards.pdf.

[27] Per the 2006 Mortgage Industry National Home Energy Rating Systems Standards, interior sheathing is not required in climate zones 1-3, although it is possible this may change to align with ENERGY STAR version 3 in future versions.

[28] Various Sources, e.g., Southern Pine Council, http://www.southernpine.com/using-southern-pine_special-topics_mold.asp.

[29] From HVI's Consumer Brochure, http://www.hvi.org/publications/pdfs/HVI_BathroomFansBrochure.pdf.

[30] For a thorough review of the topic, see Raymer, P.H., *Residential Ventilation Systems*, Somerville, MA: Building Science Press, 2010 and Rudd, Armin, *Ventilation Guide*, Somerville, MA: Building Science Press, 2011.

[31] 2012 ENERGY STAR requirements mandate energy efficient air handler motors for central-fan-integrated ventilation systems.

[32] *Practical Approaches to Residential Ventilation for Improved Durability and Indoor Air Quality*, RR 0002, Building Science Corporation, November 2000, *http://www.buildingscience.com/documents/reports/rr-0002-practical-approaches-to-residential-ventilation-for-improved-durability-and-indoor-air-quality*.

[33] *Practical Approaches to Residential Ventilation for Improved Durability and Indoor Air Quality*, RR 0002, Building Science Corporation.

[34] Rudd, Armin, *Ventilation Guide*, Somerville, MA: Building Science Press, 2011.

[35] Raymer, P.H., *Residential Ventilation Handbook*, McGraw Hill Professional, 2010.

[36] See note 34.

[37] Rutkowski, Hank, *ACCA Manual J Residential Load Calculation* (Version 8), Arlington, VA: Air Conditioning Contractors of America.

[38] Sherman, M., and I. Walker, *Right-sizing HVAC: It Can be Just Plain Wrong*, ASHRAE Journal, June 2011.

[39] Rutkowski, Hank, ACCA Manual D—Duct Design 3rd Edition, Arlington, VA: Air Conditioning Contractors of America, Arlington, VA, 2009.

[40] ACCA Manual T—Air Distribution Basics, Arlington, VA: Air Conditioning Contractors of America, 2009.

[41] ACCA Standard 5 (*HVAC Quality Installation Specification*), Arlington, VA: Air Conditioning Contractors of America, 2010.

[42] *Your New Green Home and How to Take Care of It: Homeowner Education Manual Template*. National Association of Home Builders. Washington: NAHB BuilderBooks, 2011.

[43] In February 2012, there was one manufacturer of residential-sized units, http://www.climatewell.com, but no U.S. distributor.

[44] See note 42.

[45] WaterSense estimates that overwatering accounts for half of outdoor water usage.

[46] From Table US12, Consumption by Energy End Uses. *2005 Residential Energy Consumption Survey: Household Consumption and Expenditures Tables*, Washington: U.S. Department of Energy, Energy Information Administration, 2005.

[47] Newport Partners, LLC, *Energy, Environmental, and Economic Analysis of Residential Water Heating Systems*, Washington: Propane Education and Research Council, 2010.

[48] Hendron, R., C. Engebrecht, *Building America Research Benchmark Definition*, Golden, CO: National Renewable Energy Laboratory, updated December 2009. Adapted from Equation 22.

[49] *2011 Annual Energy Outlook*, Washington: U.S. Energy Information Administration, 2011.

[50] Leslie, R.P., *Builders Guide to Home Lighting*, Troy, NY: Lighting Research Center, 1995 (http://www.lrc.rpi.edu/programs/lightingTransformation/residentialLighting/buildersguide/introduction.asp).

[51] Leslie, R.P., and K.M. Conway, *The Lighting Pattern Book for Homes*, 2nd edition, Troy, NY: Rensselaer Polytechnic Institute, 2007, http://www.lrc.rpi.edu/resources/publications/lpbh/010Preface.pdf.

[52] Conversely, efficient lights produce less heat in winter and, therefore, increase heating load. However, it costs less to use a high-efficiency heating system than to use lights to produce heat.

[53] Foldbjerg, P., S. Ipsen and N. Roy, The Impact of Window Configuration on the Performance of Energy and Daylight in Residential Buildings, Indoor Air 2011, June 5–10 2011, Austin, TX.

[54] Per cycle water usage for ENERGY STAR labeled dishwashers ranges from 1.56 gallons to 5.8 gallons.

[55] *Life Cycle Assessment of Systems for the Management and Disposal of Food Waste*, 2011, PE Americas, (Executive Summary, http://www.insinkerator.com/lca-food-waste/LCA-Executive-Summary.pdf).

[56] See *Your New Green Home and How to Take Care of It: Homeowner Education Manual Template*. Washington: NAHB BuilderBooks, 2011.

[57] U.S. Energy Information Administration, 2010 Annual Energy Outlook. Table 4 from 2012 Annual Energy Outlook Early Release review, 23 January 2012. Report Number: DOE/EIA-0383ER(2012). Accessed from http://www.eia.gov/forecasts/aeo/er/tables_ref.cfm.

[58] U.S. Energy Information Administration, Residential Energy Consumption Survey 2009, Special Report on Consumer Electronics. Accessed from http://38.96.246.204/consumption/residential/reports/electronics.cfm.

[59] Ecos Consulting, Final Field Research Report, 2006. Accessed from **http://**www.efficientproducts.org/documents/Plug_Loads_CA_Field_Research_Report_Ecos_2006.pdf.

[60] B. Neenan and J. Robinson, *Residential Energy Use Feedback: A Research*

Synthesis and Economic Framework, EPRI-1016844, Palo Alto, CA: Electric Power Research Institute, December 2008.

[61] The Energy Information Administration's Residential Energy Consumption Survey lumps appliances and consumer electronics into the same category. Because major appliances have become more efficient, most of this increase is probably from consumer electronics and plug loads.

[62] *Small Wind Electric Systems: A U.S. Consumer's Guide*, Washington: U.S. Department of Energy, Office of Energy Efficiency and Renewable Energy, http://www.windpoweringamerica.gov/pdfs/small_wind/small_wind_guide.pdf, and Carrier website dB rating for 17 SEER air-conditioning unit.

[63] Migliore, P., J. van Dam, and A. Huskey, *Acoustic Tests of Small Wind Turbines (Preprint)*. NREL/CP-500-34662, for presentation at 2004 Wind Energy Symposium, Reno, Nevada, 5-8 January 2004.

[64] Lawrence Berkeley National Laboratory's Residential Commissioning website, http://epb.lbl.gov/commissioning/index.html.

[65] Such as Building for Environmental and Economic Sustainability (BEES) software, available from http://www.nist.gov/el/economics/BEESSoftware.cfm.

[66] Information about the National Housing Quality program is at http://www.nahbrc.com/builder/quality.

[67] Leonard, D., and J. Taggart, 2010, *Final Report of Quality Assurance Activities for New Homes, Upper Marlboro, MD: NAHB Research Center*, June 2010. http://www.toolbase.org/PDF/BestPractices/AppendixAEconomicsofQuality.pdf.

[68] See scopes of work at http://www.toolbase.org (click "Best Practices" and "Scopes of Work") and at http://www.ibacos.com (click "Resources" and "Publications").

[69] As formulated into an international standard, ISO 9000.

[70] Lukachko, A., *Quality Assurance Roadmap forHigh Performance Residential*

Buildings, Somerville, MA: Building Science Corporation, 2008. To find the report, search for its title on the Building America program document portal, http://www1.eere.energy.gov/library.

[71] Building America Quality Control checklist is available from http://www.buildingscience.com/documents/guides-and-manuals/gm-building-america-quality-control-checklist/view.

[72] *ACCA Standard 9 HVAC Quality Installation Verification Protocols*, 2011, Arlington, VA: Air Conditioning Contractors of America, 2011. Accessed from https://www.acca.org/Files/?id=539.

[73] Including FMA/AAMA 100-07, *Standard Practice for the Installation of Windows with Flanges or Mounting Fins in Wood Frame Construction*, FMA/AAMA 200-09, *Standard Practice for the Installation of Windows with Frontal Flanges for Surface Barrier Masonry Construction for Extreme Wind/Water Conditions, and* AAMA 2400-10, *Standard Practice for Installation of Windows with an Exterior Flush Fin Over an Existing Window Frame*, available from http://www.aamanet.org/.

[74] Quality Built, "Nation's leading risk management company releases top builder defect data for construction industry: Quality Built Data shows builder top risk issues are preventable," Media Kit, International Builders' Show, Orlando, Fla., January 11, 2006.

[75] NAHB Research Center, "National Housing Quality Program Achieving Positive Business Results for Certified Builders and Trades," press release February 7, 2007. NAHB Research Center.

[76] Leonard, D., and J. Taggart, *Final Report of Quality Assurance Activities for New Homes (Interim Project Report) Appendix A: The Economics of Quality*, NAHB Research Center, June 2010.

[77] Adapted from, NAHB Research Center, *"Eliminating 'Hotspot' Problem Areas Key in Producing High performance Homes,"* August 22, 2011.

[78] J.D. Power and Associates, *Satisfaction with New-Home Builders and New-Home Quality Reach Historic Highs, as Home Builders Respond to Tough Market Conditions by Improving Products and Service*, press release, September 15, 2010.

[79] Levinson, Mitch, *Internet Marketing: The Key to Increased Home Sales*, Washington: NAHB BuilderBooks, 2012, 15–17.

[80] Scott, Charlie, "Solving the spec-home customer satisfaction dilemma," HousingZone.com, August 7, 2011, http://www.housingzone.com/customer-satisfaction/solving-spec-home-customer-satisfaction-dilemma.

[81] Smith, Carol, *Customer Service for Home Builders*, Washington: NAHB BuilderBooks, 2003.

[82] *See note 42.*

[83] As defined in the *2012 International Residential Code for One- and Two-Family Dwellings*, Country Club Hills, IL: International Code Council, 2012.

INDEX

recycling, 53, 54, 70, 134–35, 175

R-value, 18, 39–40, 45, 51, 54, 58, 61, 66–69, 171

scopes of work, 19, 28, 86, 143–52, 174, 177

Seasonal Energy Efficiency Ratings (SEER), 89–91, 177

site design, 5–6, 15

skylight, vi, 106, 109, 111–12

   tubular, 17, 106, 109, 111

slab-edge insulation, 33, 37

slab foundation, 31, 33, 35, 143–44

Small Wind Certification Council, i, 124

soil survey, 12, 15

solar, vii, 1, 8, 13, 16–18, 46–55, 84, 90, 98–102, 105, 110, 119–20, 135, 166–71

solar fraction, 101, 121, 171

solar heat gain coefficient (SHGC), 17, 55, 57, 84, 171

steel framing, 44, 174

steep slope, 11–12, 15–16, 49, 172, 174

storm water, 14, 19, 24, 29–30, 53, 171, 173

structural insulated panels (SIPs), 44, 45, 172

sun-tempered design, 13, 16–17, 172

surface runoff, 28

tankless water heater, 97–98, 101, 172

thermosiphoning, 120, 172

topsoil, v, 19, 23, 25, 28, 31

transfer grille, 85

U-factor, 54–55, 57, 84–85, 111, 172, 175

unvented attic design, 51–52

urea formaldehyde, 43, 136

vapor permeable, 36, 47, 50, 59, 63, 70, 171

vegetated roofs, 30, 46, 53

vegetated swales, 14, 173

ventilation, vi, 3, 17, 21, 36, 47–48, 51, 57–58, 66, 73–82, 91, 111, 128–29, 137, 146, 166–168, 170

   balanced, 77–82

   energy recovery, 76–77, 129, 167–68

   exhaust-only, 78–80

   heat recovery, 76–77, 168

   supply-only, 77–78, 82

visible transmittance, 55

walk-throughs, 160

water heater, vi, 97–102, 119

   gas, 102

   heat pump, 101–2

WaterSense, 97–98, 101, 104, 176

wetlands, 11–13, 168

wind power, vii, 124–25

xeriscaping, 95, 173

zero energy, 160